Books are to be returned on or before
the last date below.

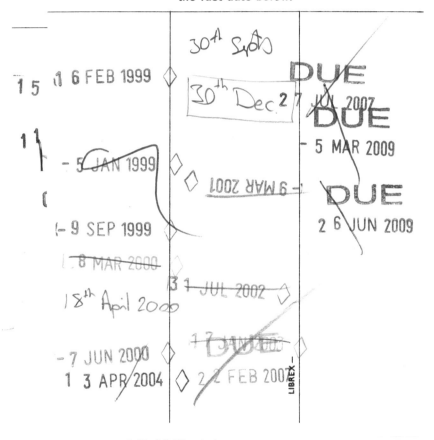

*Automatic Mesh
Generation*

Automatic Mesh Generation

Application to Finite Element Methods

P. L. George
Director of Research at INRIA
France

JOHN WILEY & SONS
Chichester • New York • Brisbane • Toronto • Singapore

MASSON   Paris   Milan   Barcelona   Bonn   1991

Copyright © Masson, Paris, 1991

The French edition of this book is published in the series *Recherches en Mathématiques Appliquées* edited by P.G. Ciarlet and J.L. Lions.

All rights reserved.

No part of this book may be reproduced by any means, or transmitted, or translated into a machine language without the written permission of the publisher.

Wiley Editorial Offices

John Wiley & Sons Ltd
Baffins Lane, Chichester
West Sussex PO19 1UD, England

John Wiley & Sons, Inc., 605 Third Avenue,
New York, NY 10158-0012, USA

Jacaranda Wiley Ltd, G.P.O. Box 859, Brisbane,
Queensland 4001, Australia

John Wiley & Sons (Canada) Ltd, 22 Worcester Road,
Rexdale, Ontario M9W 1L1, Canada

John Wiley & Sons (SEA) Pte Ltd, 37 Jalan Pemimpin 05-04,
Block B, Union Industrial Building, Singapore 2057

ISBN 0-471-93097-0 (Wiley)
ISBN 2-225-82564-5 (Masson)

A catalogue record for this book is available from the British Library

Printed in France

Abstract Automatic Mesh Generation

The finite element method is used more and more for numerical simulation of a large variety of physical problems. The first step consists of constructing an appropriate mesh of the domain under consideration. This is an important part of this simulation (in particular w.r.t. time) and proves to be delicate when complex geometries are considered. Furthermore, the quality of the numerical results is strongly linked to the quality of the corresponding mesh.

This book intends to investigate these mesh generation problems and, in particular the following four aspects:

1. *The definition of the set of information composing a mesh in order to apply the finite element method;*

2. *The development of a general methodology to conceive a mesh. In this respect a top-down analysis of the problem and associated bottom-up generation will be introduced;*

3. *The description of mesh generation algorithms for two-dimensional and three-dimensional geometries;*

4. *The presentation of usual transformation methods and their numerous applications.*

Each section is illustrated by various examples and figures corresponding to the two-dimensional and three-dimensional cases. INRIA[1], in particular with regard to the finite element code Modulef[2], offers a number of software packages dedicated to mesh creation and manipulation; most of the examples given in this book are a result of these codes.

---

[1] Institut de Recherche en Informatique et en Automatique, Domaine de Voluceau, B.P. 105 - 78153 Le Chesnay Cedex (France).
[2] Modulef: modular library for finite element simulation.

Contents

1	**General introduction**	**1**
	1.1 An outline of the finite element method	2
	1.2 General notions relative to meshes	5
	1.3 An outline of the chapters	11
2	**General mesh description**	**13**
	2.1 Introduction	13
	2.2 Constitutive information	14
	2.3 Internal representation	15
	2.4 Physical numbers	26
3	**General methodology**	**29**
	3.1 Introduction	29
	3.2 Some popular mesh generation methods	30
	3.3 Top-down analysis of the geometry	34
	3.4 The relevant attributes	38
	3.5 Minimal primal sub-sets	44
	3.6 Creating a mesh (bottom-up construction)	45
4	**Primal sets (the data)**	**51**
	4.1 Introduction	51
	4.2 Bottom-up data (construction)	52
	4.3 Bottom-up data (interest)	53
	4.4 Definitions of the basic items	55
	4.5 Summary of the bottom-up construction	58
	4.6 User data and mesh generator data	58
	4.7 An example of bottom-up data	60
5	**Manual and semi-automatic methods**	**65**
	5.1 Introduction	65
	5.2 Manual method	65

	5.3 Hexahedral topologies	70
	5.4 Cylindrical topologies	72

6 Transport-mapping methods — 77
   6.1 Introduction — 77
   6.2 General principle of the method — 77
   6.3 Method based on the domain vertices — 79
   6.4 Method preserving a polygonal contour — 82
   6.5 Transport and deformation — 90
   6.6 Data description — 92
   6.7 Application examples — 94

7 Mesh creation by solving a P.D.E. — 97
   7.1 Introduction — 97
   7.2 General principle of the method — 98
   7.3 Elliptical methods — 99
   7.4 Other methods — 106
   7.5 Some remarks — 108

8 Grid superposition-deformation — 111
   8.1 Introduction — 111
   8.2 Generation using a regular grid — 112
   8.3 Quadtree and octree — 114
   8.4 Generation based on a quadtree — 115
   8.5 Generation based on an octree — 119

9 Multiblock methods — 123
   9.1 Introduction — 123
   9.2 Multiblock methods in 2 dimensions — 124
   9.3 Multiblock methods in 3 dimensions — 127

10 Advancing front methods — 137
   10.1 Introduction — 137
   10.2 Advancing front method in 2 dimensions — 138
   10.3 Advancing front method in 3 dimensions — 147
   10.4 Extensions — 154

11 Voronoï methods and extensions — 159
   11.1 Introduction — 159
   11.2 Outline of the Delaunay-Voronoï method — 160
   11.3 Mesh generation application — 164
   11.4 The two-dimensional case — 166

11.5 The three-dimensional case 173
11.6 Extensions............................... 182

12 Creation of surface meshes   187
12.1 Introduction............................. 187
12.2 Transport-mapping method 188
12.3 Multiblock method 189
12.4 Automatic methods 190
12.5 C.A.D. approach 190
12.6 Mesh generation 208

13 Mesh transformations   215
13.1 Introduction............................. 215
13.2 Usual geometrical modifications 216
13.3 Pasting together of two meshes 220
13.4 Specified internal points, lines or faces 222
13.5 Local or global refinements 226
13.6 Mesh regularization 234
13.7 Adaptation 236
13.8 Modifying the physical attributes 242
13.9 Generation of non-vertex nodes 243
13.10 Renumbering 245
13.11 Controls and graphics 254

14 Some mesh generation packages   263
14.1 Introduction............................. 263
14.2 The Modulef library (mesh generation part) .. 263
14.3 EMC^2, a two-dimensional mesh editor 272
14.4 Various other mesh generation packages 280

15 Extensions   283
15.1 Introduction............................. 283
15.2 Data structures 283
15.3 Interpreted functions 296
15.4 Two examples of mesh construction 298
15.5 Future trends and conclusion 310

Bibliography 315
Index 331

Acknowledgments

A large number of people have contributed, in different ways, to make this book possible.

I am specially indebted to Michel Bernadou, Jean François Bourgat, François Hermeline and Pascal Joly who read the entire manuscript in detail. Their remarks and suggestions resulted in significant improvements.

I am also indebted to all the people who introduced me to numerical analysis and, in this way, guide me to the domain of mesh generation, a fascinating field of investigation! Among them, Philippe Ciarlet, Roland Glowinski, Olivier Pironneau, Alain Perronnet and Pierre Arnault Raviart should be mentioned specially.

I am equally grateful to all the people who helped me by providing certain illustrations incorporated in this book, in particular: B. Palmerio, J.P. Dehette, J. Derampe, J.A. Desideri, C. Doursat, J.J. Droux, P. Germain-Lacour, A. Golgolab, A. Hassim, F. Hecht, Ph. Hidden, P. Joly, A. Marrocco, J. Periaux, F. Pistre, E. Saltel, M. Vénéré and M. Vidrascu.

I would also like to thank Maryse Desnous who did all the figures and schemes which were not performed automatically.

Finally, I would like to thank Philippe Ciarlet and Jacques-Louis Lions who have accepted this work in the *Recherches en Mathématiques Appliquées* collection.

This text is translated from "Génération Automatique de Maillages. Applications aux Méthodes d'Eléments Finis", RMA no. 16, first published by Masson, Paris, at the end of 1990. The first translation, done by the author, has been carefully read by R. George and H. Du Toit who made significant changes to clarify the translation.

Chapter 1

General introduction

The simulation of various physical phenomena (in chemistry, thermal analysis, electromagnetism, mechanics of solids, fluid mechanics, etc.) can be written in terms of partial derivative equations which can be solved numerically using the finite element method.

The essence of this method consists of calculating approximate values of the solutions desired (temperatures, stresses, pressure, velocity, magnetic field, etc.). These values are computed at some points in the domain of interest, called the nodes. From this set of values, it is possible to derive the solution values at any position; this computational step is based on the use of chosen interpolation functions. The above numerical calculation requires, as a first step, the construction of a mesh of the domain where the problem is posed in order to define the nodes.

This phase of preprocessing is very important in the sense that the generation of a valid mesh in a domain with a complex geometry is not a trivial operation and can be very expensive in terms of the time required. On the other hand, it is crucial to create a mesh which is well adapted to the physical properties of the problem under consideration, as the quality of the computed solution is strongly related to the quality of the mesh.

Firstly, this chapter introduces the basic ideas of the finite element method. Then, the notion of a mesh is introduced with the relevant definitions and notations. Finally, the different chapters of this book are briefly described.

1.1 An outline of the finite element method

The finite element method was first conceived and used by engineers in the early fifties. Among the first references are [Argyris-1954-1955] and [Clough-1960]. Due to the improvement of computers during the following years, this method became more and more popular for numerical simulation of a large range of physical problems written in terms of partial derivative equations. The implementation of the method, as well as the definition of most of the classical finite elements, took place in the sixties and seventies [Zienkiewicz-1971], [Oden-1972], [Zienkiewicz-1973] ... Numerous books and papers devoted to this topic have appeared since. Both the technical and mathematical aspects were investigated. Among the main references, we can mention, for example, [Babuska,Aziz-1972], [Strang,Fix-1973], [Ciarlet-1978], [Bathe-1982], [Oden,Carey-1984], [Hughes-1988] and [Argyris,Mlejnek-1986-1988] and, in addition, for certain types of applications [Raviart,Thomas-1983], [Glowinski-1984] and [Johnson-1987].

As a tentative presentation of the basic principles of the finite element method, we choose a very simple example: *the membrane problem*. Let Ω be a domain, assumed to be polygonal, with boundary Γ; this problem can be written as follows:

$$\begin{cases} -\Delta u = p \text{ in } \Omega \\ u = 0 \text{ on } \Gamma \end{cases} \quad (1.1)$$

For this problem, the vertical displacements, u, are to be computed at any point located in the membrane, assumed planar initially, which is clamped along boundary Γ and is subject to a vertical force of density p.

A better adapted formulation, called the *variational formulation*, or *weak formulation*, is derived from the above formulation written *in terms of partial derivative equations* (P.D.E.), and can be written as follows:

$$\begin{cases} \text{Find } u \in V \text{ such that} \\ a(u,v) = f(v), \forall v \in V \end{cases} \quad (1.2)$$

with

$$\begin{cases} a(u,v) = \int_\Omega \left(\frac{\partial u}{\partial x}\frac{\partial v}{\partial x} + \frac{\partial u}{\partial y}\frac{\partial v}{\partial y} \right) dxdy \\ f(v) = \int_\Omega pv\,dxdy \end{cases} \quad (1.3)$$

where V is the space of admissible displacements.

1.1. AN OUTLINE OF THE FINITE ELEMENT METHOD

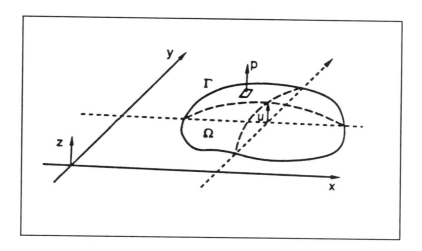

Figure 1.1: *Membrane problem.*

This formulation has the following two major advantages:

it is closer to the physical aspect of the problem: the form $a(u,v)$ represents the deformation energy of the system and $f(v)$ represents the potential energy of the external forces applied to the material;

it is mathematically suitable: an appropriate selection of space V ensures the existence and uniqueness of the solution, u.

This **continuous problem** is then replaced by an **approximated problem**. In practice, the explicit solution of the continuous problem is in general not possible. This leads us to investigate **approximate solutions** using the finite element method. It consists of the construction of a sub-space V_h, which is a **finite dimensional** sub-space of space V, and the definition of u_h, an approximate solution of u, where u_h is the solution of the following problem:

$$\begin{cases} \text{Find } u_h \in V_h \text{ such that} \\ a(u_h, v_h) = f(v_h), \forall v_h \in V_h \end{cases} \quad (1.4)$$

It can be shown that this problem has a unique solution, u_h, and that the **convergence** of u_h to the solution, u, is directly related to the manner in which the functions v_h of space V_h approach the functions v of space V, and therefore the manner in which space V_h is defined.

In this respect, the finite element method consists of the construction of a space V_h of finite dimension such that, on one hand, suitable approximation properties are obtained and, on the other hand, it is convenient from the point of view of computer implementation. This construction is based on the three following basic ideas:

1. the creation of a mesh, denoted by $\mathcal{T}_h$, of domain Ω. The domain is written as the finite union of elements K, as will be described later (cf. definition 1.1);

2. the definition of V_h as the set of functions v_h, whose restriction to each element K in $\mathcal{T}_h$ is a polynomial;

3. the existence of a basis for the space V_h whose functions have a *small support*.

Thus, the restriction of a function u_h to any element K is written as:

$$u_h = \sum_{i=1}^{N} u_i v_i$$

where N denotes the *number of degrees of freedom* of element K, u_i is the value of the function at the degrees of freedom and v_i is the basis function i of the polynomial space defined previously. A finite element is then characterized by a suitable choice of (following the notation used in [Ciarlet-1978]):

- K the geometrical element (triangle, quadrilateral, etc.);

- Σ_K the set of degrees of freedom defined on K. These degrees are the values at the **nodes** of the function (Lagrange type finite element), or those values and directional derivatives of the function (Hermite type finite element);

- P_K the basis polynomials on K.

The **implementation** of a Finite Element Method can be summarized as follows:

- the mathematical analysis of the problem with, more specifically, its related variational formulation and the investigation of the properties of the latter;

- the construction of a triangulation (the mesh) of the domain under consideration, i.e., the creation of $\mathcal{T}_h$;

- the definition of the different finite elements, i.e., the choice of (K, P_K, Σ_K);

- the generation of element matrices due to the contribution of each element, K, to the matrix and right hand side of the system;

- the assembly of the global system;

- the consideration of *essential*[1] boundary conditions (in this case, $u = 0$ on Γ);

- the solution of the system, i.e., the computation of the solution field approaching the sought solution;

- the post-processing of results.

This scheme is quite general and, in fact, so is that of any problem solved by a finite element method.

The purpose of this book is to describe phase 2 of this scheme assuming the other phases. One may observe that, while numerous papers deal with this topic, there exists no comprehensive book, at least to the author's knowledge, except for some particular classes of applications.

1.2 General notions relative to meshes

A mesh of a domain, Ω, is defined by a set, $\mathcal{T}_h$, consisting of a finite number of segments in one dimension, segments, triangles and quadrilaterals in two dimensions and the above elements, tetrahedra, pentahedra and hexahedra in three dimensions. The elements, K, of such a mesh must satisfy a certain number of properties which will be introduced later. The first concerns the *conformity*, with respect to the following definition:

Definition 1.1 : $\mathcal{T}_h$ is a *conformal* mesh of Ω if the following conditions hold:

1. $\overline{\Omega} = \cup_{K \in \mathcal{T}_h} K$

2. all elements K of $\mathcal{T}_h$ have a non-empty interior

---
[1] The nature of boundary conditions is connected to the chosen variational formulation.

3. the intersection of 2 elements in $\mathcal{T}_h$ is such that it is either:

- reduced to an empty set,
- reduced to a point,
- reduced to an edge,
- reduced to a face.   □

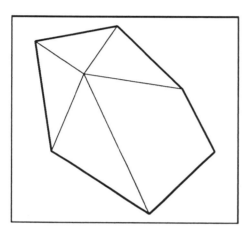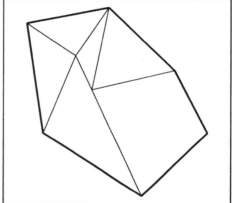

Figure 1.2: *Conformal and non-conformal meshes.*

This definition implies that $\mathcal{T}_h$ covers Ω in a *conformal* manner (figure 1.2 left hand side); it corresponds only to the geometrical aspects.

In practice, $\mathcal{T}_h$ is the partitioning of Ω, as accurately as possible. When Ω is not a polygonal (polyhedral) domain, $\mathcal{T}_h$ will only be an approximate partitioning of the domain.

Finite elements are associated with the basic geometrical elements of the mesh. This operation consists of the definition of nodes and interpolation functions (polynomials for example).

Some finite element packages can consider elements with shapes different from those mentioned above (for example, the pyramid with a square base).

Furthermore, for some methods of computation, the conformity is not globally satisfied at the mesh construction level. The related continuity is

1.2. GENERAL NOTIONS RELATIVE TO MESHES

then ensured by an appropriate definition of the finite element nodes and the associated interpolation functions.

The elements constituting the mesh must generally satisfy some specific properties. In this respect, we have the following:

- Properties of a geometric nature:

    - The variation in size between two adjacent elements must be progressive and discontinuity from elements to elements must not be too stiff.
    - The density of elements in some regions of the domain must be higher (for example, in case of stiff gradients in the solution).
    - When elements are of a triangular type, the existence of a non-obtuse angle is to be avoided.

        Definition 1.2 : A mesh consisting of triangular elements is said to be *non-obtuse* if they do not have obtuse angles. □

        More generally, the elements, K, must be sufficiently regular and must satisfy certain properties relative to their shape (flattening shape, etc.).

    - Elements must satisfy anisotropic features (i.e. some directions are to be preferred).
    - ...

- Properties of a physical nature:

    These are strongly connected to the physical aspects of the problem under consideration. They can be achieved if geometrical properties of the above type are present. In fact, one has to produce relatively thin elements, isotropic or anisotropic elements, elements with specified shapes, etc. according to the physical behaviour of the problem.

More generally, the theorems for the error estimation in the finite element method [Ciarlet-1978] are globally related to the ratio $Q = \frac{\rho}{h}$, where ρ is the radius of the inscribed circle (sphere) and h the diameter of the element considered (the reason for h in the notation $\mathcal{T}_h$). The abstract error estimate is due to the asymptotic behaviour of expressions based on the quantity Q^{-1} when all elements in the mesh are considered. From a mathematical point of view, the convergence of a finite element method is based on the analysis of such quantities as h tends to 0. However, in practice, h is limited by $max_K h_K$ and remains "small".

Numerous algorithms for the construction of meshes for two-dimensional and three-dimensional geometries exist. The choice of the method will be strongly connected to the geometry of the domain under consideration.

First of all, one has to note that there are two main classes of methods. To be more precise, let us give the following definitions:

Definition 1.3 : The *connectivity* of a mesh is the definition of the connection between its vertices. □

Definition 1.4 : A mesh is called *structured* if its connectivity is of the finite difference type. □

and on the other hand

Definition 1.5 : A mesh is called *unstructured* if its connectivity is of any other type. □

The first class of methods considers only *structured meshes* or *grids*, while the second class considers *unstructured* meshes. For a structured mesh, the connectivity between nodes is of the type (i,j,k), i.e., assuming the indices of a given node, the node with indices (i,j,k) has the node with indices $((i-1),j,k)$ as its left neighbour and that with indices $((i+1),j,k)$, etc) as its right neighbour; this kind of approach is convenient for geometries for which such properties are suitable, i.e., for generalized quadrilateral or hexahedral configurations; for different situations, specific treatment is necessary to conceive this type of structure.

In order to be as general as possible, the structured case will only be described briefly whereas the unstructured case will be considered in more detail (more precisely, the case of structured meshes will not be explicitly developed).

Considering the different mesh generation algorithms, it is convenient to classify them into seven classes:

1. "Manual" or "semi-automatic" methods suitable for cases with relatively simple geometries,

2. Methods which construct the mesh of the real domain as a mapping, by a suitable transformation, of a mesh of simple geometry,

3. Methods based on the solution of a system of partial derivative equations,

4. Methods based on the deformation and local modification of an easily obtainable grid,

1.2. GENERAL NOTIONS RELATIVE TO MESHES

5. Methods based on the composition of meshes of sub-sets of the domain, obtained by the methods in case 2 or 3,

6. Methods deriving the final mesh, element by element, from the boundary data,

7. Methods using a composition of meshes of sub-sets based on geometrical or topological modification of these meshes, obtained by any of the previous methods.

Remark 1.1 : This classification is necessarily more or less arbitrary; however, while it is not unique, it includes the different possible approaches. □

Remark 1.2 : The type of method to use depends on the geometry of the domain and the manner in which it is described. □

For some situations, a *modeller* type approach is possible. It consists of a boolean description and construction of the domain in terms of unions, intersections, etc. of geometrically elementary shapes. Each of these elementary shapes is then meshed and the final mesh is given by the union of these sub-meshes.

Each of these classes of methods are briefly described below:

- **Class 1** : in this case, we have enumerative methods (points, edges, faces and elements composing the mesh are given explicitly) and methods suitable for particular geometrical situations (hexahedral, cylindrical shape) which use the specific properties of the geometry explicitly (where the connectivity is known in advance).

- **Class 2** : thanks to the facility of a mapping function, the mesh of a reference domain is mapped onto the real domain.

- **Class 3** : in a sense, this approach looks like that seen in class 2 above, but now the mapping function is not given initially but is computed by solving partial derivative equations in order to satisfy certain useful properties (element density, orthogonality, etc.).

- **Class 4** : the main application in this class is of type "quadtree" for dimension 2 and "octree" for dimension 3. The domain is enclosed in a quadrilateral or quadrilateral parallelepipedon which is split into boxes. These boxes are constructed by decomposition based on a quaternary tree (dimension 2) or octal tree (dimension 3). This grid is then used to create the desired mesh.

- **Class 5** : the preceding classes 2 and 3 can also be used in block decomposition methods. The domain is discretized roughly by a set of blocks of elementary shape, after which the methods of class 2 or 3 are applied to each block.

- **Class 6** : there are two approaches: advancing-front methods (frontal methods) and algorithms based on the Voronoï-Delaunay's construction. These methods create internal points and elements, starting from the boundary of the domain. This boundary is given either in a global manner (for example an analytical definition), or in a discrete manner (as a list of the edges or triangular faces approximating it).

- **Class 7** : in this case, the problem is split into a set of sub-problems of simpler complexity. These are then solved by one or several of the previous classes of methods and the final result is obtained by transformations and pasting together of partial results (class 5 is a particular case of this more general approach).

The above summary of the different mesh generation methods emphasizes some problems among which are:

- The multiplicity of algorithms;

- The greater or lesser generality of each method; some are performant or simply applicable to particular geometries;

- The large variety of data to be provided (points, lines, surfaces, boundaries, coarse meshes, previously created meshes, etc.), the quantity and consequently the manner in which it is provided;

- The necessity for an adequate definition of the notion of a mesh in such a way that it is convenient in terms of future computation. As a consequence, one has to specify the contents of a mesh precisely (i.e., choose the pertinent values to be stored) and define the access to the corresponding information (i.e. the way of extracting a value).

For these reasons, generally speaking, it is advisable to define the *notion* of a mesh precisely. Having done this, the choice of a *general method* for the conception of these meshes is fundamental. Thanks to this type of method, all cases, even those of complex geometry, can be treated. The analysis of the mesh generation problem is a consequence of this choice. Moreover, the manner in which the different methods are *implemented* on computer is strongly connected to the preceding options.

1.3 An outline of the chapters

This book aims to present some solutions for the effective treatment of these problems. Thus, we introduce a possible definition of the notion of meshes, its structure and a general methodology for the conception of meshes. The algorithms for the creation, both for the two-dimensional ($2D$) and the three-dimensional ($3D$) case, will be described and, in addition, the case of surfaces ($2D\frac{1}{2}$) will be discussed. Tools for the modification and exploitation of meshes will then be detailed.

Numerous application examples will illustrate the different problems encountered. In this respect, software developed at INRIA and, in particular, that of the Modulef library [Bernadou et al. 1988] will be presented. In addition, other mesh generation software packages will also be mentioned.

In short:

- Chapter 2 gives an acceptable **definition** of a mesh and, in addition, some possible mesh **structures** are mentioned.

- In chapter 3, a brief survey of the different mesh generation methods is given after which a **methodology** of a general application for the conception of meshes is introduced. The notion of **primal sets**, i.e., only those sets which need to be considered effectively in order to obtain a mesh of any domain Ω, is introduced.

- In chapter 4, the manner in which these primal sets can be meshed is explained. A **formal process** for the conception of meshes is given. It details how to use the mesh generation algorithms and indicates the method of providing the relevant data.

- The following chapters are devoted to the different **approaches possible**. Cases of simple geometry or with a topology known in advance are described in chapter 5. In chapter 6, the methods based on the transformation of simple meshes by a pre-defined mapping function are discussed. The creation of meshes by solving partial derivative equations is dealt with in chapter 7. Grid superposition and deformation type methods are mentioned in chapter 8. Methods based on multiblock decomposition are developed in chapter 9. Advancing front techniques are detailed in chapter 10. Mesh generators based on the Delaunay-Voronoï method are discussed in chapter 11.

- In chapter 12, the main lines of argument for and ideas of methods suitable for surface meshes are introduced.

- A collection of **tools** for geometrical or topological modifications of existing meshes is presented in chapter 13.

- In chapter 14, a brief overview of the capabilities of the Modulef library regarding mesh creation and manipulation is given. Moreover, other software packages for mesh generation are listed.

- Lastly, some additions of technical interest and some specific applications are given in chapter 15, concluding with a summary of future developments and possible extensions.

This book concludes with an index and a bibliography.

Chapter 2

General mesh description

2.1 Introduction

According to the applications (thermal, structural, mechanical, fluid, electromagnetic, etc. problems), numerical simulation by the finite element method requires the mesh data of the domain under consideration. The latter can be reduced to an object, a set of objects or to the above items with a zone surrounding them, or to the exterior of one or several objects.

Whatever the case may be, the *mesh* must contain all useful information when considering the different steps in the numerical computation (geometry, definition of loads, computation of matrices, solution of systems, visualization of results, etc.).

This information includes the following types:

- **Geometrical information** : this consists of:

    - the description of the mesh, i.e., the geometrical covering-up of the domain of computation as accurately as possible. This covering-up is formally composed of the set of elements.
    - the description of all the information previously used at the construction of the elements, i.e., the *history*.

- **Information about the finite element interpolation**: the mesh must be compatible with the chosen interpolation in order to be able to define it (type of interpolation, number of nodes per element, the list of these nodes and their location).

- **Physical information**: the values contained in the mesh must permit the definition of physical characteristics of materials, applied loads, sources of temperature, etc. as well as the non-natural boundary conditions (locations and imposed values; the distinction between the natural and non-natural boundary conditions depends on the type of finite element selected).

2.2 Constitutive information

To ensure these objectives, the internal representation of a mesh must allow one to find any value (quantity stored in this representation) quickly or to easily compute any value of interest from the information contained (quantity derived from information stored in this representation). A mesh is therefore, in general, a set of values and tables of values containing the list of geometrical elements composing the mesh. For each element, the following information must be known:

- the *nature* of the element: segment, triangle, quadrilateral, tetrahedron, pentahedron, hexahedron or particular element (see below);

- the *history* of the element: an element is composed of faces, edges and vertices; these *items* have been generated from data which must be accessible (for example, which vertex of the element is located on a curve of a given equation);

- the list of the *vertices* of the element;

- the *connectivity* and *topology* of the element; the connectivity (cf. definition 1.3) describes the liaison between the vertices, the topology is defined as follows:

  Definition 2.1 : The *topology* of an element is the description of its edges and faces by its vertices.   □

- the *number* and the list of its *nodes*;

  Definition 2.2 : A *node* is a point supporting one or several unknowns.   □

- the *coordinates* of its vertices;

- the *physical attributes* of the element:

  - a *sub-domain or material number*. Assigned to each element, this number allows the characterization of this element with respect to an operation of the following type:

2.3. INTERNAL REPRESENTATION

$$\boxed{\text{sub-domain number} \rightarrow \text{"volumetric" processing}}$$

— *reference numbers* or *references*. Assigned to items composing an element (faces, edges, points), this number allows the characterization of these items with respect to an operation of the following type:

$$\boxed{\text{reference number} \rightarrow \text{"sub-volumetric" processing}}$$

The precise effect of these two kinds of attributes will be detailed in section 2.4 and in chapter 3.

The previous description contains information of the three types mentioned above (geometrical point of view, definition of the interpolation and physical description, cf. 2.1).

Remark 2.1 : The nodes and the vertices may coincide or not. □

Remark 2.2 : The mesh procedure can include the creation of nodes necessary for the interpolation step. The concept of the node (connected to that of interpolation) is introduced from the mesh generation step (a priori geometrical notion only) to permit the optimal numbering of nodes with respect to the solution method envisaged. In practice, when creating a mesh, it is sufficient to know the number and the relative position of the nodes without further information (for example, it suffices to know that there is one node per edge, the fact that it is at the midpoint of the edge or not is of no importance). □

Remark 2.3 : To describe the physical aspect of the items (faces, edges, points) of a mesh, the notion of a reference number is sometimes reduced to that of a boolean flag (i.e., a value of type *yes-no* or *0-1*) which indicates if a special operation is to be done. □

2.3 Internal representation

The meshes created by the different mesh packages contain information of the previously described types. However, the effective storage differs from one package to another (in the absence of a unique norm accepted by all! concerning the normalization problem of the storage of descriptive values associated with a mesh, one can look at the norms SET [norme-c], IGES [norme-a] and PDES [norme-b]); it is clear that the selection of an organization for the structure of values constituting a mesh does not lead to a unique result. This choice is governed by the use one wants to make of this structure. In this respect, it is easy to see that operations as different as visualization, computation of element matrices, solution by direct, iterative or multigrid techniques are strongly connected to the way in which the values are accessible. As a consequence, it is not easy to define a universal structure which is optimal in every respect.

To illustrate this data organization problem, let us consider the solution selected in the Modulef library [Bernadou et al. 1988] for mesh storage: the definition of a *data structure* (D.S. below) of type *NOPO*[1]; described below, it is the storage method of all meshes created by the Modulef library. One can extract the desired information from this structure and, depending on the purpose, define a different storage system which would be better adapted to the problem under consideration.

In itself, the D.S. *NOPO* is well adapted to common computations of the finite element method, moreover this structure minimizes the quantity of values to be stored: it corresponds to a sequential access, element by element, and clearly assumes the case of unstructured meshes. For mesh visualization purposes, it permits the creation of a different data structure of type *GEOM*[2]. The latter is more suitable because of the way the values are accessed but, on the other hand, it requires more memory.

In the case of structured meshes and for certain particular applications, some of the information in the above list becomes implicit; as a consequence, the contents and the organization of the mesh structure will be simplified:

- case of structured meshes: such a mesh is defined by two parameters: n the number of points along one direction (the x-axis) and p that corresponding to the y-axis (we consider the two-dimensional case). Element i of the grid, constructed using this data, has as vertices the points with numbers $i+1, i+2, i+2+n, i+1+n$. It is therefore not strictly necessary to store these values as they are implicitly known by the number of each element.

- case of meshes containing only simplices[3]: in this case, the vertices of the elements can be known by the data for each vertex, a, of the oriented list $(a_i, i = 1, 2, ...)$ of its neighbours. Let us consider the two-dimensional case, with vertex a and the following list:

---

[1] NOPO: generic name of the mesh data structure in the Modulef library.
[2] GEOM: generic name of a different structure of values associated with a mesh.
[3] simplex: triangle in dimension 2, tetrahedron in dimension 3.

2.3. INTERNAL REPRESENTATION

- neighbours of a : $a_1, a_2, a_3, ...$

Then the triangles with point a as a vertex are the following:

- $(a, a_1, a_2), (a, a_2, a_3), ...$

The reader can refer to [Rivara-1986] for an application utilizing this storage method for the purpose of refinement and derefinement of meshes consisting only of simplices.

The main idea is in fact the **definition of an organization** and the development of various **tools to access** values easily.

2.3.1 Brief description of D.S. NOPO

Constructed according to the same principles as the other data structures in this package [Modulef et al. 1989], D.S. *NOPO* of the Modulef library consists of six tables of pre-defined organization.

1. Table *NOP0* : General information.

2. Table *NOP1* : Description of optional supplementary tables.

3. Table *NOP2* : General mesh description.

4. Table *NOP3* : Optional pointer.

5. Table *NOP4* : Vertex coordinates.

6. Table *NOP5* : Sequential element description.

The full description of this D.S. is listed in chapter 15.

2.3.2 Brief description of D.S. GEOM

A brief description of the *GEOM* data structure of the Modulef library is given below. It is important to note the conceptual difference between the organization of this structure and that of the previous one.

Structure *GEOM* consists of eight tables of pre-defined organization.

1. Table *GEO0* : General information.

2. Table *GEO1* : Description of optional supplementary tables.

3. Table *GEO2* : General mesh description.

4. Table *GEO3* : Description of elements in terms of faces.

5. Table *GEO4* : Description of faces in terms of edges.

6. Table *GEO5* : Description of edges in terms of vertices.

7. Table *GEO6* : Vertex coordinates.

8. Table *GEO7* : Vertex reference numbers.

A full description of this D.S. is given in chapter 15.

2.3.3 Manipulation of the mesh D.S.

As previously mentioned, the efficiency of a package depends very much on how easily its data structures are accessed. To illustrate this point, let us consider the data structures *NOPO* and *GEOM* of the Modulef library.

The set of tables constituting D.S. *NOPO* and *GEOM* can be in core (C.M.) or be stored in a sequential file.

Mesh creation modules produce a D.S. *NOPO*, manipulation modules use this data structure as input and produce a new structure of the same type as a result; in this respect, data structure *NOPO* is the **natural interface** which enables the communication between modules.

Graphic modules to visualize meshes or corresponding solutions use either data structure *NOPO* in dimension 2, or data structure *NOPO* or *GEOM* in dimension 3, as input, or some derived data structures in the case where the interpolation is "finer", i.e., the nodes are not only located at the element vertices.

In addition, and for particular applications, one can find a set of programs which enables the access and manipulation of values contained in the D.S.s. The reader is referred to chapter 15 where a list of these programs is given along with a description of their functions.

Thus, the part of the Modulef library devoted to mesh purposes is not only a set of programs for mesh creation, but forms a complete environment consisting of:

- The definition of a data structure;

- Various mesh creation modules;

- Modules for mesh modification and visualization;

- Technical tools for the easy manipulation of the D.S. associated with the mesh and for the manipulation of stored values;

2.3. INTERNAL REPRESENTATION

- Moreover, as seen below, this set of "capabilities" is closely connected to a methodology for the conception of meshes, which will be described fully in chapters 3 and 4.

The efficiency of any package (in general) results from this kind of organization. It is, at least, the minimum requirement for the conception of a consistent mesh generation package.

2.3.4 Connectivity and topology of a triangle

The numbering of vertices, edges and faces of elements is pre-defined in such a way that some properties are implicitly induced. In particular, the oriented numbering of vertices enables us to compute the surface of a triangle with a positive, or directional, sense. It also allows us to evaluate, for each edge, directional normals.

In the case of a triangle (with connectivity (1)(2)(3)), the first vertex (1) having been chosen, the numbering of the others is deduced through anticlockwise rotation (figure 2.1). The edges are then determined by:

- edge [1] : it runs from vertex (1) to vertex (2) and runs in the direction (1) to (2),

- edge [2] : (2) $\rightarrow$ (3)

- edge [3] : (3) $\rightarrow$ (1)

From this connectivity, the topology (definition 2.1) can now be defined. The topology of all geometric elements is listed in 2.3.6.

In the case of a triangle, one can also find a different convention for the numbering of the edges. The edge [i] is the edge which is opposite vertex (i). While it makes simplified notation possible, this definition cannot be applied to other types of elements.

2.3.5 Local numbering of nodes

The optional intermediary nodes located on the edges are numbered sequentially in the order of edges following nodes at vertices, the edges running in the defined direction. Then the nodes on the faces are numbered and finally the internal nodes are enumerated.

To state this numbering convention precisely, we reconsider the case of a triangle and provide four examples of some well-known finite elements:

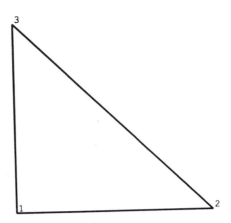

Figure 2.1: *Topology of a triangle.*

1. Lagrange type finite element of degree 1, or a three-noded triangle. This element only has nodes at the vertices, as in figure 2.1.

2. Finite element of the type primal hybrid. This element has nodes at the midpoints of its edges, and the vertices are not nodes. The first node is that of the first edge, the others are numbered in the order of the edges (figure 2.2).

3. Lagrange type finite element of degree 2, or six-noded triangle. The nodes of this element are the vertices and the midpoints of its edges. The first three nodes are the three vertices, the next three are those located on its edges considered in the order stipulated previously (figure 2.3, left hand side).

4. Lagrange type finite element of degree 3. For this element, the nodes are:
    - its vertices;
    - two points per edge;
    - one internal point.

The vertex nodes are numbered first, according to the local numbering of these vertices, then the nodes on the edges are listed according to their rank and finally the internal node is numbered (figure 2.3, right

2.3. INTERNAL REPRESENTATION

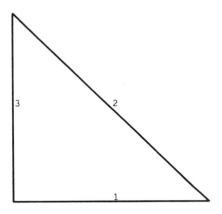

Figure 2.2: *The primal hybrid "P1" Lagrange triangle.*

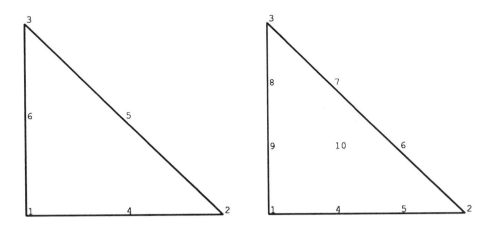

Figure 2.3: *The "P2" triangle and "P3" triangle.*

hand side). In the case of an element with several internal nodes, one must take care to number them in a logical hierarchy, for example, with respect to their position).

2.3.6 Topology and numbering of nodes (general list)

The topological definition of usual elements and the local numbering of their nodes are given in this section. The convention followed is that of the Modulef library (it is, of course, possible to define a different convention). Having chosen the first vertex, one has:

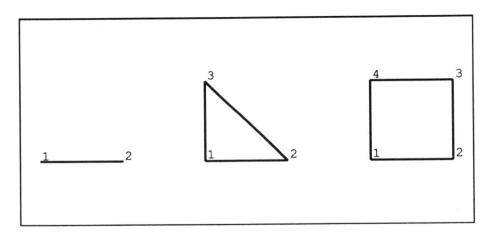

Figure 2.4: *Segment, triangle and quadrilateral.*

1. The segment :

    - $(1) \to (2)$

2. The triangle : numbering in an anticlockwise direction (in dimension 2; in the case of dimension 3, the surface has no sign).

    - edge [1] : $(1) \to (2)$ (nodes 4, 5, ... if there are any supplementary nodes)
    - edge [2] : $(2) \to (3)$ (...)
    - edge [3] : $(3) \to (1)$

3. The quadrilateral : numbering in the same manner (see remark for a triangle)

2.3. INTERNAL REPRESENTATION

- edge [1] : (1) → (2) (nodes 5, 6, ...)
- edge [2] : (2) → (3) (...)
- edge [3] : (3) → (4)
- edge [4] : (4) → (1)

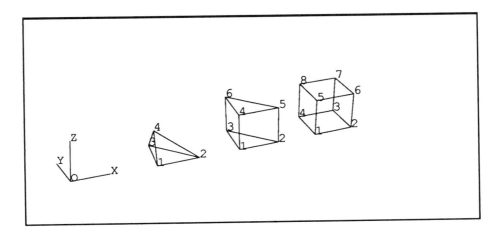

Figure 2.5: *Tetrahedron, pentahedron and hexahedron.*

4. The tetrahedron : trihedron $(\vec{12}, \vec{13}, \vec{14})$ assumed positive. ($\vec{ij}$ denotes the vector with origin point i and extremity point j).

- edge [1] : (1) → (2) (nodes 5, ...)
- edge [2] : (2) → (3) (...)
- edge [3] : (3) → (1)
- edge [4] : (1) → (4)
- edge [5] : (2) → (4)
- edge [6] : (3) → (4) (nodes ..., p)

Any face, seen from the exterior, is in the positive direction:

- face [1] : (1) (3) (2) (nodes $p + 1$, ...)
- face [2] : (1) (4) (3) (...)
- face [3] : (1) (2) (4)
- face [4] : (2) (3) (4)

5. The pentahedron : trihedron ($\vec{12}, \vec{13}, \vec{14}$) assumed positive
    - edge [1] : (1) → (2) (nodes 7, ...)
    - edge [2] : (2) → (3) (...)
    - edge [3] : (3) → (1)
    - edge [4] : (1) → (4)
    - edge [5] : (2) → (5)
    - edge [6] : (3) → (6)
    - edge [7] : (4) → (5)
    - edge [8] : (5) → (6)
    - edge [9] : (6) → (4) (nodes ..., p)

   Any face, seen from the exterior, is in the positive direction:
    - face [1] : (1) (3) (2) (nodes $p+1$, ...)
    - face [2] : (1) (4) (6) (3)
    - face [3] : (1) (2) (5) (4)
    - face [4] : (4) (5) (6)
    - face [5] : (2) (3) (5) (6)

6. The hexahedron : trihedron ($\vec{12}, \vec{14}, \vec{15}$) assumed positive
    - edge [1] : (1) → (2) (nodes 9, ...)
    - edge [2] : (2) → (3) (...)
    - edge [3] : (3) → (4)
    - edge [4] : (4) → (1)
    - edge [5] : (1) → (5)
    - edge [6] : (2) → (6)
    - edge [7] : (3) → (7)
    - edge [8] : (4) → (8)
    - edge [9] : (5) → (6)
    - edge [10] : (6) → (7)
    - edge [11] : (7) → (8)
    - edge [12] : (8) → (5) (nodes ..., p)

   Any face, seen from the exterior, is in the positive direction:

2.3. INTERNAL REPRESENTATION

- face [1] : (1) (4) (3) (2) (nodes $p+1$, ...)
- face [2] : (1) (5) (8) (4)
- face [3] : (1) (2) (6) (5)
- face [4] : (5) (6) (7) (8)
- face [5] : (2) (3) (7) (6)
- face [6] : (3) (4) (8) (7)

To this list of elements corresponding to the usual geometric shapes, one can, in addition, define other elements developed for specific problems; we can mention, for example, the elements of junction which simulate a hinge (i.e. two segments for the junction between two beams or two quadrilaterals in the R^3 space for the junction between two planes.

Remark 2.4 : If necessary, new elements can be developed. It is necessary to define them in a consistent manner in this case. □

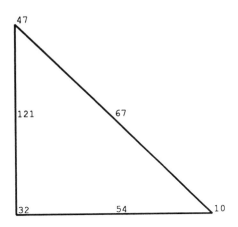

Figure 2.6: *Global and local numbering.*

The topology and the local numbering of nodes establishes the correspondence between the global numbering of nodes and their local positions. For example, in the case of the triangle shown in figure 2.6, if we have the following list of global node numbers: 32 10 47 54 67 121 , and if the element is of type "P2", then it is easy to deduce that node 32 is the first vertex, ..., and that node 54 is located on edge 1 of the element, etc.

Thus, there is no ambiguity in the numbering.

2.4 Physical numbers

The tables describing the mesh contain information (cf. sections 2.1 and 2.2) which enables us to select elements, faces, edges and nodes with the goal of assigning physical values. These attributes, the *sub-domain number* and the *reference numbers* will be very useful in future computational steps.

The *sub-domain number* associated with each element makes a *global processing* of elements sharing the same number possible. A correspondence will be established between this value and the physical material characteristics to which the element belongs (conductivity, heat capacity coefficient, Young's modulus, Poisson's coefficient, etc.) or the loads to which it is subject (distribution of volumetric strains in dimension 3, distribution of surface strains in dimension 2), or finally, for any specific treatment which is required.

Reference numbers, or simply *references*, are associated with each item constituting an element (faces, edges and nodes). They make *global processing* of all items sharing the same number possible. A correspondence will be established between this value and:

- the loads to which the item is subject (coefficient of exchange, flow through a surface or distribution of surface forces in dimension 3, flow across a line or line forces in dimension 2). In such cases, the value characterizes a face or an edge.

- the assignment of non-natural boundary conditions (of imposed value type) at nodes. In this case, it corresponds to a value characterizing a node.

- finally, any specific operation which is required.

In particular, the *reference* can be used to indicate a geometrical operation (projection, etc.) to be executed during the mesh generation step or the post-processing of it. Thanks to this number, one can indicate that a point, an edge or a face has been defined on a surface, a curve, etc. when elaborating the mesh data (the *history* of the item is thus accessible).

Conventionally, the numbering of sub-domains and references starts at the value 1; however, for references, the value 0 will indicate that no special treatment is foreseen.

Table 2.1 shows these two kinds of attributes schematically, which can be logically classified by type:

2.4. PHYSICAL NUMBERS

type	dimension 3	dimension 2	dimension 1
$\mathcal{V}$	sub-domain	-	-
$\mathcal{S}$	reference	sub-domain	-
$\mathcal{L}$	reference	reference	sub-domain
$\mathcal{P}$	reference	reference	reference

Table 2.1 : *Logical organization of attributes.*

In table 2.1, $\mathcal{V}$, $\mathcal{S}$, $\mathcal{L}$ and $\mathcal{P}$ denote the set of volumetric, surface, line and point information respectively.

This organization consists of classifying information by type; this type of classification will provide numerous advantages, for example, when calculating the element matrices.

Chapter 3

General methodology

3.1 Introduction

A detailed study of the different mesh creation methods is given in chapter 5 and the following chapters. In order to establish a general methodology for the conception of meshes, the present chapter is devoted to the description of the main ideas of the most popular of these methods.

The conception of a mesh can be decomposed into three steps:

- The analysis of the problem;

- The formal definition of the mesh generation process;

- The actual construction with, firstly, the definition of the data and secondly, the actual creation of the mesh.

The first step consists of an analysis of the geometry of the domain and of the physical problem to be solved; this analysis will be processed in a **top-down manner** (i.e. the decomposition of a complex problem into a series of simpler sub-problems).

The formal construction of the mesh, which constitutes the second step, takes the results of the top-down analysis into account and is based on a **bottom-up construction** (i.e. the definition of simple objects making the solution of the entire problem possible, step by step).

The actual creation of meshes of these different objects, which is the third step, is achieved through the use of appropriate mesh generation algorithms, and consists of two phases:

- a phase consisting of the **creation of the relevant data**;

- a phase corresponding to the **actual generation** of meshes.

We must note that the mesh obtained will consist only of elements defined by their vertices as modules for the creation of meshes and modules for the transformation of meshes only consider such meshes; so, when the mesh of a domain is obtained, one has to enrich it if finite elements with non-vertex nodes are desired. On the other hand, the numbering, both of elements or of nodes, is not optimal or specific (except for particular cases) with respect to the solution method selected later (the methods for renumbering will be described in chapter 13).

3.2 Some popular mesh generation methods

The methods which are available for the realization of meshes in dimension 2, $2\frac{1}{2}$[1] and 3 (or $2D$, $2\frac{1}{2}D$ and $3D$ meshes) can be classified according to classes 1 to 7 discussed in chapter 1. The following enumeration provides, for each of these classes, some possible approaches. The possible application, its limits and the nature of the data to be provided are discussed for each case.

Class 1 :

- M1 : "**Manual generation**" by the definition of all useful information. In this case, the user defines the elements by their vertices. This approach is quite suitable for domains with very simple geometries or for those which can be covered by a limited number of elements. The resulting mesh will often be post-processed by transformations (in particular splitting) to obtain, finally, more accurate and precise meshes.

- M2 : "**Semi-automatic generation**" by the definition of all useful information for the construction of a mesh assuming a previously known connectivity. A simple mesh is used as a basis of a more complex construction. This approach is suitable for domains with hexahedral or cylindrical geometries: from a two-dimensional mesh of a section, the corresponding volumetric elements are defined layer by layer with the help of a given transformation.

---
[1]this notation is employed for the surfaces in the space R^3.

3.2. SOME POPULAR MESH GENERATION METHODS

Class 2 :

- M3 : Generation of a mesh by **transport-mapping** of a reference mesh. This corresponds to a domain with an elementary geometry (triangle, quadrilateral, etc.). The mapping function is predefined to ensure certain properties (respecting contours, etc.). In this case, the data is a discretization of the contour of the domain.

Class 3 :

- M4 : Generation of a mesh by the **explicit solution of partial derivative equations** formulated on a reference mesh. It corresponds to a domain with an elementary geometry in which a structured grid is developed easily (quadrilateral, hexahedron). A mapping function is defined to ensure some required properties (respecting of contours, orthogonality of elements, variable density of elements, etc.). In this case, the domain is also defined via its contour.

Class 4 :

- M5 : Application of techniques of **overlapping** and **deformation** of simple meshes in such a way that the real domain is covered accurately. The domain, defined by its contour, is included in a grid which is composed of quadrilateral or hexahedral boxes, constructed in order to satisfy some properties (the number, the size and the repartitioning of these boxes must be in agreement with the problem under consideration). This grid is then deformed in such a way that the real domain is covered accurately, after which the boxes of this grid are split, if desired, to obtain the elements of the final mesh. A priori suitable for any geometry, this method can be difficult to implement in some situations.

Class 5 :

- M6 : Generation of the final mesh by **structured partitioning** of a coarse mesh composed of blocks, assuming a simple geometric shape. In this approach, a coarse partitioning of the domain composed of blocks of simple elementary geometry (segments, triangles, quadrilaterals, tetrahedra, etc.) is given. The interfaces between these blocks must be defined carefully, on one hand, to ensure the validity of the result and, on the other hand, to determine the nature of the partitioning to which they will be subject (number of subdivisions, etc.) precisely. This technique enables the consideration of a domain with an arbitrary shape; nevertheless, in the case of a complex shape, one has to define a large number of initial blocks.

Class 6 :

- M7 : Generation of a mesh of a domain from its boundary by an **advancing-front method**. The boundary is defined either in a global or discrete way as a polygonal approximation of the contour in the form of segments in dimension 2 and, in dimension 3, as a polyhedral approximation in the form of triangular faces. A front, initialized by the contour, is established. From an edge (a face) of the front, one or several internal elements are created, the front is then updated by suppressing or adding some edges (faces), the process iterating as long as the front is not empty. This technique is suitable for the creation of triangular or tetrahedral meshes of arbitrary domains; note that the implementation of such a method must be designed carefully to ensure the convergence of the process.

- M8 : Generation of the mesh of a domain from the points of its boundary by an approach of type **Delaunay-Voronoï**. The boundary is defined as for method M7. This type of mesh generator usually carries this name as it relies on the Delaunay-Voronoï construction which results in the creation of a mesh of a convex hull of a set of points; nowadays, this approach seems to have the most general application.

- M9 : Generation of a mesh from a set of points located in the domain or describing its contours. This application is a particular case of method M8.

Class 7 :

- M10 : Generation of meshes by **combining** previously created meshes. This approach relies on the use of geometrical transformations (symmetry, for example) or topological transformations

3.2. SOME POPULAR MESH GENERATION METHODS

(global refinement, for example). In this case, the geometrical properties of the domain are considered fully in order to mesh only the useful parts; the mesh of the various parts, which can be derived from others (by symmetry, rotation, etc.), is then obtained using simple transformations and the final mesh is simply the pasting together of all the different sub-meshes.

The main types of data necessary when implementing methods M1 to M10 have diverse natures of which we can enumerate the following types:

- the enumeration of all elements from the list of vertices;
- the contour or boundary data;
- functions for the transformation of existing meshes;
- coarse meshes;
- meshes with simpler complexity (for example two-dimensional meshes, serving as a support for the construction of three-dimensional meshes).

It is convenient to follow the tree-structure in terms of volumes, surfaces, lines and points as introduced in the previous chapter.

As the notion of a mesh is identified with that of a volume (surface in dimension 2), we obtain the following classification:

- volumes are defined by surfaces;
- surfaces are defined by contours;
- contours are defined by lines;
- lines are defined by points.

Thus the notion of a point is the most elementary notion of the process of construction.

The associated scheme is then:

$$\mathcal{V} \Leftarrow \mathcal{S} \Leftarrow \mathcal{L} \Leftarrow \mathcal{P}$$

Having this schematic view of the capabilities of mesh generators, we will analyze the domain and the related physical problem below in order to propose a general mesh construction methodology. This analysis must take the purely geometrical features of the problem and, in addition, its physical aspects into account.

3.3 Top-down analysis of the geometry

The purpose of this step is to best utilize the geometrical features of the domain in the view of minimizing the work necessary to generate a mesh, as well as making this step more reliable.

The method investigated is a *top-down* method and leads to splitting the problem under consideration (possibly a complex one from a geometrical point of view) into a series of sub-problems which are easier to solve or which are better adapted with regard to the capabilities of the available algorithms.

This step has three objectives:

- To minimize the mesh generation operation and make it more reliable by considering the possible repetitive features present in the domain. In this respect, it is often useless to create the mesh of the entire domain if some parts of the domain can easily be obtained from others.

- To adapt the region under consideration to the capabilities or robustness of the mesh generator. Any shape can be considered successfully by a "powerful" mesh generator but one has to split the domain into adapted parts when a "poor" mesh generator is used.

- To obtain a mesh enjoying some special features, for example:
    - to create quadrilaterals for some regions of the space;
    - to obtain a varying density of the elements;
    - to impose a given point, a given line, ... in a mesh;
    - to obtain symmetry in the resulting mesh, as this symmetry is not produced directly by the mesh generator used; in this case, only a part of the domain is processed and the user orders the symmetry himself.

In most cases, a domain can be subdivided into geometrically simple sub-domains. Various possibilities exist to obtain this partitioning:

1. The geometrical repetitive features (existence of symmetries, translations, rotations, etc.) are recorded to define the sub-sets, called *primal sub-sets*, which will be effectively considered for mesh generation opposed to the sub-sets, called *secondary sub-sets*, which can be derived with the help of usual transformations (symmetries, translations, rotations, etc.).

3.3. TOP-DOWN ANALYSIS OF THE GEOMETRY

Figure 3.1 and following figures illustrate this idea. The entire domain (figure 3.1) is such that we only have to consider the domain shown in figure 3.2, the primal sub-set.

Thus, the two following definitions are introduced:

Definition 3.1 : A *primal sub-set* of a domain is a sub-set of that domain whose mesh is obtained by a mesh algorithm (refer also to definition 3.3 which refines this notion). □

Definition 3.2 : A *secondary sub-set* of a domain is a sub-set of that domain whose mesh is derived by transformation of a mesh of another sub-set. □

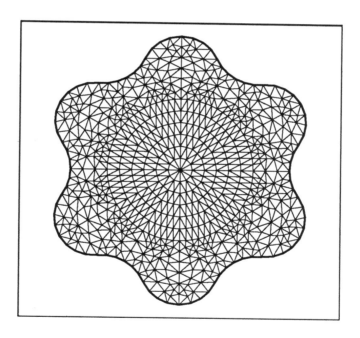

Figure 3.1: *The entire domain (after creation of its mesh)*.

2. If such possibilities do not exist, one can split the domain artificially into geometrically simpler regions which form the primal sub-sets. Figure 3.3 illustrates this feature where a splitting line is artificially defined.

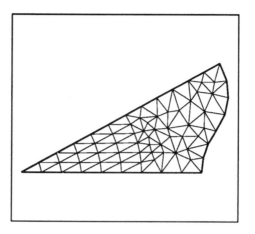

Figure 3.2: *The only sub-set to be considered (its mesh).*

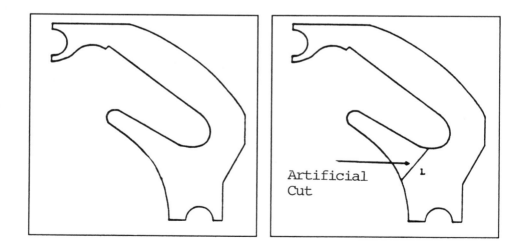

Figure 3.3: *The domain of interest and a possible partitioning.*

3.3. TOP-DOWN ANALYSIS OF THE GEOMETRY

Line L (figure 3.3) is introduced to split the domain into two primal sub-sets of simpler geometries.

3. Another way of achieving this simplification is to obtain different meshes which are only slightly different, by creating only one of them, considered as a primal sub-set, and creating the others by simple modification of that one. Figure 3.4 illustrates this situation.

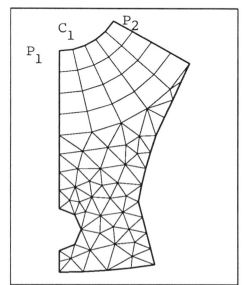
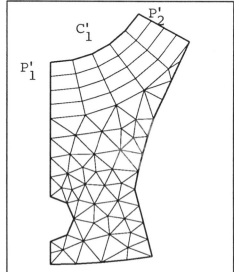

Figure 3.4: *Meshes deductible from one another.*

The mesh of the domain on the right-hand side (figure 3.4) is generated in the same way as the one on the left-hand side by simple modification of points $P1$ and $P2$ and the definition of curve $C1$; these two domains differ only by the radius of the circle corresponding to their curves and consequently by the data of the parameters defining them.

4. Besides, the valid pasting together of meshes (to ensure conformal features) can lead to the definition of a priori non-evident primal sub-sets. Figure 3.5 illustrates this situation.

The $3D$ mesh of figure 3.5 (left part) can be derived, as will be described later, from the $2D$ mesh of its basis. It is therefore necessary

 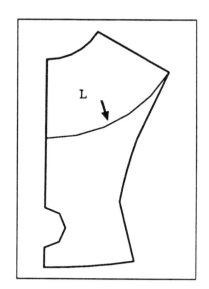

Figure 3.5: *Constraints related to the pasting of two meshes.*

to secure a line L when considering this $2D$ mesh (see figure 3.5, right-hand side) and, thus, define two primal sub-sets. The construction of the $3D$ mesh can be done easily by pasting together these two objects. This is due to the fact that the pasting region is derived from the "transport" of line L, defined in the $2D$ problem.

By combining these various approaches, we obtain the primal sub-sets which are to be considered effectively. As they have been derived exclusively from a geometrical analysis, they are called *primal geometrical sub-sets*.

In some situations, this notion must be refined, as we will see below.

3.4 The relevant attributes

In this section, we will take the influence of the physical aspects of the problem into account when considering the notion of a primal sub-set, since it is, at present, only connected to the geometrical aspects of the problem.

Due to the analysis of the physical aspects of the problem, we must consider the following:

- the different materials constituting the domain;

3.4. THE RELEVANT ATTRIBUTES

- the applied loads (volumetric, surface, line and point data);
- the imposed boundary conditions (prescribed values at nodes, etc.).

We already know that these operations, at the solution level, will be linked to the attributes (see section 2.2) characterizing the elements of the mesh, its faces, edges or nodes, according to scheme 3.6.

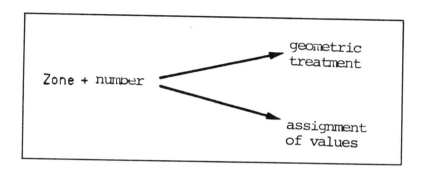

Figure 3.6: *Correspondence : attributes / operations.*

It is necessary to identify two kinds of attributes: those relative to geometrical considerations and those relative to physical considerations. Next, the set of attributes to be assigned with respect to these two objectives must be found.

3.4.1 Attributes corresponding to geometrical considerations

For the description of surfaces (in $3D$) or curves (in $2D$ or $3D$) known analytically as a function $f(x,y,z)$ or $f(x,y)$, reference numbers of faces or edges are used. These curves or surfaces can be used, for example, for the definition of the boundary or parts of it.

To clarify this point, we consider figure 3.7 which shows a domain consisting of two materials bounded by curve C joining points $P1$ and $P2$. As the mesh of the domain must include this curve or a close approximation of it, it is necessary to assign a reference number to the corresponding line to be able to identify it.

Thus, when splitting curve $(P1-P2)$ into segments, the points obtained can be generated on the curve with equation $f(x,y) = 0$ by mapping the corresponding points located on cord $(P1 - P2)$, or by using any other suitable method (figure 3.7). This process is activated when the previously

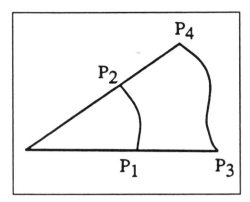 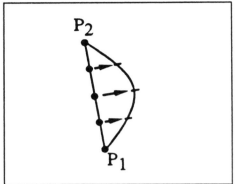

Figure 3.7: *Identifying a curve and associated partitioning.*

defined reference number is encountered. The effect of this number is thus clearly identified.

We note that two geometrical primal sub-sets have been defined in this way.

The same problem appears for boundary $(P3 - P4)$ in figure 3.7 which can be described by two analytical functions $f_1(x, y)$ and $f_2(x, y)$. In this case, we partition boundary $(P3 - P4)$ into two sub-lines in order to assign the two numbers needed for the definition of the two curves, f_1 and f_2, thus enabling the execution of the two appropriate processes.

To this end, point $P5$ has been introduced naturally on part $P3 - P4$ of the boundary.

Remark 3.1 : One can assign an attribute (reference number) to any part of a boundary, or a sub-domain number to any region of the domain, in the view of activating any specific set of instructions (graphical issues, sorting or selection of specific points, edges or faces, etc.). □

3.4.2 Attributes relative to the physical problem

The purpose of this section is to take the physical properties of the problem to be investigated into account in order to facilitate the future assignment of the physical data (material coefficients, description of applied loads, definition of non-natural boundary conditions or prescribed values at some nodes of the mesh, etc.).

With regard to their function, two types of numbers are defined:

3.4. THE RELEVANT ATTRIBUTES

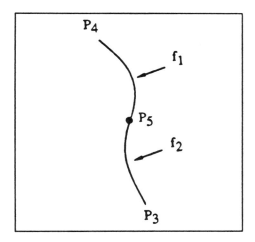

Figure 3.8: *Splitting connected with two descriptions.*

Sub-domain number

This attribute (cf. section 2.2) which is a volumetric notion, enables us, on the one hand, to assign the physical characteristics of each constitutive material and, on the other hand, to compute the loads to which these materials are subject. Generally, in the case of finite element simulation, this attribute is used to define the *integrations on the domain*, i.e., computations of type:

$$\int_\Omega f(.)dx$$

As, in practice, each mesh generator (as will be shown in chapter 5 and the following chapters) allows the assignment of one sub-domain number solely to all elements created, it is necessary to create a separate mesh for each material and then to define as many primal sub-sets as required.

Thus, figure 3.9 shows that we need to consider at least two primal sub-sets.

A sub-domain number is associated with sub-domain Ω_1, a different number is associated with sub-domain Ω_2. In this way, each element of the mesh is clearly characterized with respect to its constitutive materials.

Moreover, let us assume that the loads applied to material Ω_2 are of two types: $F_1(x, y)$ in one part of Ω_2 and $F_2(x, y)$ in the other part; if it is not possible to express F_1 and F_2 as an operation of type:

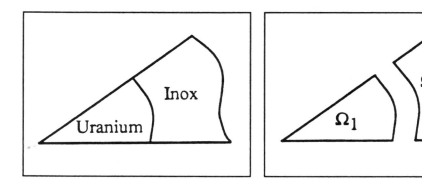

Figure 3.9: *The 2 primal sub-sets associated with the 2 materials.*

Sub-domain number ⇔ F_1 or F_2,

it is useful and convenient to define two different sub-domain numbers resulting in the following configuration:

First sub-domain number ⇔ F_1
Second sub-domain number ⇔ F_2

Finally, three primal sub-sets are encountered (figure 3.10).

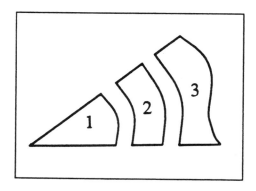

Figure 3.10: *The 3 primal sub-sets.*

In this case, the following situation occurs:

Sub-domain number 1 ⇔ material 1
Sub-domain number 2 ⇔ material 2 and loads F_1
Sub-domain number 3 ⇔ material 2 and loads F_2

3.4. THE RELEVANT ATTRIBUTES

Reference number

This attribute is defined to characterize sub-volumetric items (faces, edges and nodes in dimension 3, edges and nodes in dimension 2).

In addition to its purely geometrical role (cf. section 3.4.1), the reference number makes the relation between any part of the boundary and any specific set of processes possible; this is achieved in the same manner as for the sub-domain number. In general this attribute corresponds to *point operations* (for example non-natural boundary conditions where this is the attribute associated with nodes of the mesh which is relevant) or *integrations on the boundary* i.e., operations of type:

$$\int_\Gamma f(.)d\gamma$$

where Γ denotes a portion of this boundary. In this case, it is the edge ($2D$) or face ($3D$) reference numbers which allow the activation of this operation.

Thus, it is possible to distinguish by two distinct reference numbers a part of the boundary which is a priori uniquely definable from a geometrical viewpoint.

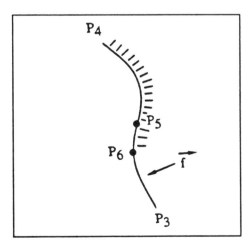

Figure 3.11: *Geometrical and physical attributes.*

To illustrate this point, we reconsider figure 3.8, assuming that boundary ($P3-P4$) is clamped on one part and subject to a load on another part (figure 3.11). As two reference numbers already exist for the geometrical definition of this boundary, it is convenient to add an additional one to be able to identify the physical data. The following scheme holds:

Reference number 1 $\Leftrightarrow$ curve $f_1(x,y)$ and prescribed conditions
Reference number 2 $\Leftrightarrow$ curve $f_2(x,y)$ and prescribed conditions
Reference number 3 $\Leftrightarrow$ curve $f_2(x,y)$ and loads

In this way, point $P6$ is introduced naturally to separate the two parts of the boundary .

3.5 Minimal primal sub-sets

The geometrical analysis led us to distinguish the geometrical primal sub-sets. The physical problem and the necessity of assigning sub-domain and reference numbers resulted in the definition of finer sub-sets, defined formally as follows:

Definition 3.3 : A *minimal primal sub-set* is a sub-set of the domain whose mesh will be created with the help of a mesh algorithm. □

With regard to definition 3.1, this notion takes the physical aspects of the problem into account. This distinction will not be made in the notation after here. These *minimal primal sub-sets*, or *generic sub-sets*, will constitute the sub-sets to be considered effectively. They will be designed in such a way that they will adapt to the capabilities of the available algorithms (see remark 3.1).

Remark 3.2 : The different types of sub-sets may coincide. □

As a consequence of these definitions, and by regarding the primal sub-sets of interest, the *lines* and *points* needed for their construction have been located (in addition, the necessary *faces* have been located in $3D$). This information clearly forms part of the data to be provided.

Each of these sub-sets and their constitutive items (points, lines and faces) possess desired attributes which make the appropriate operations possible at the geometrical level (when creating the mesh) as well as at the physical level (when computing the matrices and the right-hand-side vector of the system).

When manipulating meshes of sub-sets (see chapter 13), it is possible to modify some of their attributes; in this way, the mesh of a primal set with given attributes may result in the creation of a mesh of another geometrically "equivalent" set, whose attributes may be different. This process is shown in figure 3.12.

3.6. CREATING A MESH (BOTTOM-UP CONSTRUCTION)

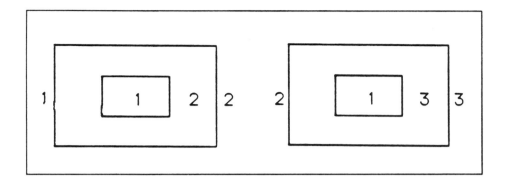

Figure 3.12: *Modifying attributes.*

We assume that the right-hand side of figure 3.12 can be derived from the left part of the same figure by translation. For physical reasons (different properties), this construction requires the following correspondence:

Sub-domain number 1 ⇔ Sub-domain number 1
Sub-domain number 2 ⇔ Sub-domain number 3
Reference number 1 ⇔ Reference number 2
Reference number 2 ⇔ Reference number 3

in the case of translation and:

Reference number 1 ⇔ Reference number 3
Reference number 2 ⇔ Reference number 2

in the case of symmetry.

It is therefore necessary to assign the appropriate attributes to the initial mesh (i.e. to define as many numbers as necessary) in accordance with the operations that will be applied to it.

3.6 Creating a mesh (bottom-up construction)

Each generic sub-set, defined following the above methodology, will be meshed separately using an appropriate mesh generator. Remember that these sub-sets result from the previous analysis (geometrical and physical), combined with the necessity of adapting each sub-set to the capabilities of the selected mesh generator.

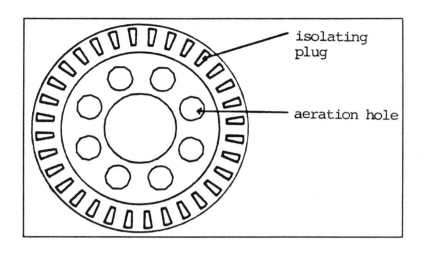

Figure 3.13: *The domain under consideration.*

Meshes of all *secondary* sub-sets are derived from those of the generic sub-sets by simple manipulation. The final mesh of the entire domain is obtained by the pasting together of these sub-meshes.

Following this philosophy, the complete mesh is generated in a *bottom-up* construction: the final covering up of the domain is derived from meshes of simple sub-sets.

A primary consequence of this methodology for the conception of meshes is the reliability of the final mesh: it is nothing more than the union of primal and secondary meshes of regions of simpler geometries. Since these are valid, the final result is valid.

To illustrate this methodology, we consider figures 3.13 and 3.15; the scheme in figure 3.14 illustrates an acceptable top-down decomposition and associated bottom-up construction (we show one possible scheme but do not claim it is the only one).

The final mesh contains a large number of elements and nodes. However, to obtain it, only one sixteenth of the domain to which the proposed methods were applied, was considered.

As an exercise, we suggest that the reader analyzes the $3D$ problem in figure 3.16 (top of the figure) in order to verify that the proposed scheme is suitable both for the problem analysis (see figure from top to bottom) as well as for the construction of the associated mesh (cf. same figure seen from bottom to top).

3.6. CREATING A MESH (BOTTOM-UP CONSTRUCTION)

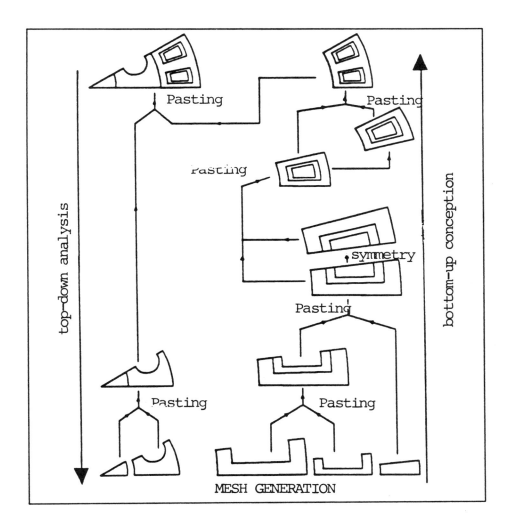

Figure 3.14: *Top-down analysis and bottom-up conception.*

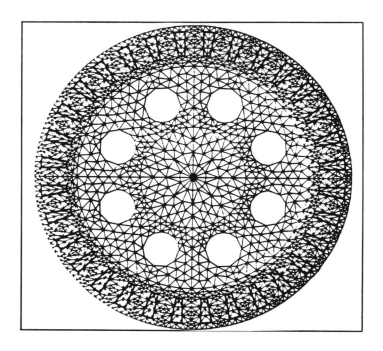

Figure 3.15: *The mesh of the entire domain.*

3.6. CREATING A MESH (BOTTOM-UP CONSTRUCTION)

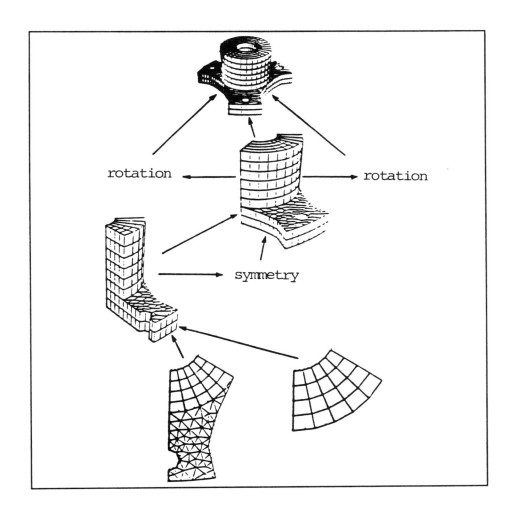

Figure 3.16: *Analysis and conception (a 3D example).*

Chapter 4

Primal sets (the data)

4.1 Introduction

In the previous chapter the (minimal) primal sub-sets, suitable for meshing, were found. Assuming that the mesh creation process is a "black box" with the following scheme:

$$primal \quad sub-set \quad + \quad Mesh \quad generator \quad \Rightarrow \quad Mesh$$

then the problem which remains to be solved corresponds to the entering of the data which is relevant to the generator. To be more precise, it corresponds to constructing the set of information used in the definition of the data. This process must be as simple as possible while allowing the most general situations to occur. In practice, there are two types of data:

- The *user* data which is composed of the set of information to be **effectively** provided by the user;

- The *mesh generator* data which will be **deduced** from the previous data, generally by some automatic process.

The nature of the data required in all the cases considered leads us to propose a formal data conception method. As it is strongly connected to the methodology seen in the previous chapter, this conception will proceed in a bottom-up manner.

The notions of points, lines, faces and volumes, the *basic data items*, will be introduced in a natural way and their interest will be discussed briefly.

4.2 Bottom-up data (construction)

Data capturing when creating a mesh is one of the difficulties of mesh generation. Among the problems connected with this operation, one can encounter the following:

- the data may vary from one mesh generator to another,

- the quantity of data to be provided must be minimal but sufficient for an adequate description of the geometric domain and physical problem,

- in particular, redundancies, which are potential error sources, must be avoided,

- finally, the user effort necessary must be simplified as much as possible.

To reach these objectives, the data definition follows a *bottom-up form*; this is one of the solutions that ensure both the reliability of the result and minimization of the amount of data to be provided.

From a geometrical point of view, mesh generation data (cf. chapter 3) consists of meshes, surfaces (meshed or not), boundaries (meshed or not), contours (meshed or not), lines (meshed or not), segments (meshed or not) and points:

- a mesh, being the result of a former process, is known via the file in which it is contained (this file is created once the process has been completed), or via a set of tables residing in core (C.M.);

- the mesh of a surface (for example, provided as the description of the boundaries of a three-dimensional domain in order to use an advancing-front or Voronoï method) will be the result of a previously created mesh (case 1 above is then encountered), or will be the output of a C.A.D.[1] package;

- a surface (not meshed) will be given in an analytical or parametrical form;

- a boundary will be described via its mesh or by describing functions;

---

[1] C.A.D.: Computer Aided Design.

- a contour, in 2 dimensions, is given in an analytical form or as a polygonal. The latter is constituted by a list of discretized lines. In 3 dimensions, a contour will be defined by a surface (see above);

- a line is described analytically or seen as a list of segments;

- a segment is defined by its two end-points and its possible partitioning;

- a point is known by its position.

As the geometrical and physical analysis of the mesh generation problem (cf. chapter 3) has been considered, a set of data related to the definition of the different primal sub-sets have been exhibited. In this respect, points, lines, etc., have been pointed out and will naturally form part of the data to be provided.

Thus, this data creation process leads us to define the *characteristic points* or *coarse points* which make the definition of the *characteristic lines* possible. From these lines, the *faces* in 3 dimensions, or *contours* in 2 dimensions are defined; faces are used in the definition of *volumes*, *coarse elements* or *meshed surfaces*.

4.3 Bottom-up data (interest)

These definitions are naturally structured following the abstract scheme shown below:

$$\mathcal{P}\text{oints} \Rightarrow \mathcal{L}\text{ines} \Rightarrow \mathcal{F}\text{aces} \Rightarrow \mathcal{V}\text{olumes}$$

- the complex elements (i.e. the volumes) are derived from simpler elements (faces deduced from lines, lines derived from points). So any modification is performed easily;

- each item (point, line, face, volume) contains numerous information (geometry and physical attributes) in a condensed form;

- the pasting together of meshes, operations which are done frequently, are reliable as common regions are well defined and assume a smaller complexity.

These features will be illustrated below using the example of a $\mathcal{L}$ine.

4.3.1 Automatic transmission of information

It is easy to emphasize this property with this example.

A $\mathcal{L}$ine is defined (cf. 4.4.2 below) by its two end-points, its directional sense, the number of points located between its end-points and their distribution, its geometry and finally its reference number. This last attribute will be assigned automatically to all intermediary points and all resulting sub-segments; in this way, only one piece of information is transmitted to a set of items.

4.3.2 Chain modifications

To modify a line (retaining this example), i.e., a set of segments and points, it is sufficient to modify its end-points (two $\mathcal{P}$oints) or its geometry (only one value) or the number of its intermediary points (one value again) or, finally, the distribution of these points (one value). Therefore a slight modification of the input produces a very different data set. This feature remains valid in the case of $\mathcal{F}$aces (see below).

4.3.3 Unique definition of items

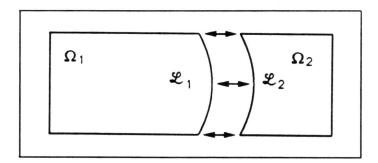

Figure 4.1: *Unique definition.*

It is obvious to see, using the same example, the interest of the concept of $\mathcal{L}$ines which is defined once and once only to obtain the pasting together of two meshes.

Let us assume (figure 4.1) that two primal sub-sets are to be pasted together: the line of pasting ($\mathcal{L}_1$ for sub-domain 1, and $\mathcal{L}_2$ for sub-domain 2) is unique. As the mesh creation for these two regions is done in one single execution by a call to the two appropriate mesh generators, we use the

4.4. DEFINITIONS OF THE BASIC ITEMS 55

same line $\mathcal{L} = \mathcal{L}_1 \equiv \mathcal{L}_2$ as data for each of them; therefore the common points are identical and the pasting together is correct.

Observe that this uniqueness of data minimizes the amount of information to be input and, in addition, ensures the reliability of the result.

It is, in fact, possible to mesh this domain in two different executions, but this would result in redundant information, which could give rise to related errors; two lines would be input: $\mathcal{L}_1$ and $\mathcal{L}_2$.

This observation remains valid for $\mathcal{F}$aces (see below).

4.4 Definitions of the basic items

The set of items relevant to the data creation associated with a domain and to the related physical problem is defined in this section.

4.4.1 Definition of a point

We understand by the definition of a point either a characteristic $\mathcal{P}$oint (i.e. a point which is explicitly provided by the user) or a point generated on the contour (on its lines), or finally, all points created by the mesh generator inside the domain. Whatever the case, the attributes of a point are:

- its coordinates;

- its reference number.

Characteristic $\mathcal{P}$oints are defined by the user. These $\mathcal{P}$oints must be as numerous as is necessary to define the domain properly; i.e., from these points, the definition of all characteristic $\mathcal{L}$ines, $\mathcal{F}$aces, etc. is possible.

The points, characteristic (points effectively given) or not (for example points created automatically on the contours), constitute the set of points describing the domain before mesh creation.

Definition 4.1 : Two points are called *geometrically identical* if their geometrical attributes (their coordinates) are the same; they are *completely identical* if all their attributes are the same (i.e., their coordinates and reference number). □

4.4.2 Definition of a line

The characteristic $\mathcal{L}$ines are input by the user and serve, on one hand, to define the geometry and the physical problem (thanks to the attributes) and, on the other hand, to construct more complex items.

A ℒine is described by a set of points located on a "curve" joining its two end-points which are characteristic 𝒫oints. Consecutive points on a line define straight edges so that the line is approximated by a *polygonal line*; a ℒine is described by the following information:

- its 2 end-points;

- its directional sense of definition;

- the number of points to be created between its end-points and their distribution (geometrical or arithmetical progression or manual description);

- its geometry (straight line, identified curve (circle, ellipse, etc.), curve with a given equation, curve defined by splines, etc. or a set of piecewise lines when all intermediary points are given explicitly;

- its reference number.

Recall that the reference number can be used either to define the geometry or to assign physical values to be associated to line items. In fact, the attribute assigned to the line is transmitted to all its undefined items (i.e., to all its points except the end-points which are characteristic 𝒫oints (and therefore are already known), and to all the segments joining its points, two by two).

Definition 4.2 : Two lines are called *geometrically identical* if their geometric attributes (end-points, number and position of points, etc.) are the same; they are *completely identical* if all their attributes are the same (except the directional sense). □

From its geometrical description (cf. above), the line is approximated by a set of segments. In the case where the line is not simply defined by the enumeration of its segments, they need to be created. According to the definition provided, this construction can be based upon:

1. an orthogonal projection (cf. figure 4.2, left-hand side);

2. an angular projection (cf. figure 4.2, right-hand side);

3. an evaluation;

4. and so on.

4.4. DEFINITIONS OF THE BASIC ITEMS

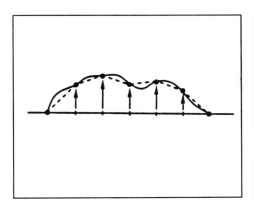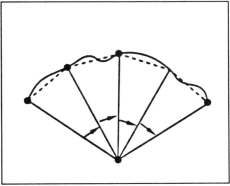

Figure 4.2: *Approximating forms of a line.*

In the first case, the cord joining the two end-points of line is split into sub-segments respecting a given distribution. These points are then mapped on the real curve to obtain the discretized form of the line. This approximation is then used later on.

In the second case, a centre of projection is given and angular sectors are constructed (according to the given distribution), where their intersections with the line define its discretized form.

The third case implies that the intermediate points on the line under consideration are known via the evaluation of given functions with the corresponding parameter values.

For other types of line definition, the discrete form is obtained differently (see for example, in chapter 12, the case where a definition based on splines is used).

4.4.3 Definition of a face

In addition to its geometrical definition from, for example, characteristic $\mathcal{L}$ines, a face has a reference number (in $3D$) to allow for all global treatment envisaged later on.

This attribute will be assigned automatically to all items of the face not yet defined; they correspond to the points, edges and sub-faces which are not characteristic $\mathcal{P}$oints or characteristic $\mathcal{L}$ines or included in such items.

The remark in section 4.4.2. remains valid for $\mathcal{F}$aces.

Note that the concept of a $\mathcal{F}$ace does not exist in dimension 2 (see remark 4.3).

The mesh of faces (dimension 3) is based on the above definition and may proceed in different ways, described in the following chapters.

4.4.4 Definition of a volume

$\mathcal{V}$olumes are defined by the $\mathcal{F}$aces constituting them. Each volume has a sub-domain number which is assigned to all elements resulting from its mesh.

Remark 4.1 : In dimension 2, the concept of a $\mathcal{V}$olume is replaced by that of a $\mathcal{S}$urface which possesses, in the same way, a sub-domain number. □

The mesh of volumes (dimension 3) or surfaces (dimension 2) is based on their definition and proceeds following different methods, described in the following chapters.

4.5 Summary of the bottom-up construction

The creation of a mesh can be summarized as the splitting of the domain into primal sub-sets and then considering only the sub-sets derived from the minimal ones. For each of these regions, one must provide the set of characteristic $\mathcal{P}$oints from which the set of characteristic $\mathcal{L}$ines, $\mathcal{F}$aces and $\mathcal{V}$olumes is derived. From this information, the mesh generator generates the finite elements (defined by their vertices as nodes) covering the domain.

In practice, the mesh creation process is often preceded by a preprocessing phase which generates the data necessary for the mesh generator from the user data.

4.6 User data and mesh generator data

This phase of preprocessing proceeds from the user data and constructs the data adapted to the mesh generator envisaged, and is therefore strongly linked to it. This step is crucial: it must be automatized and controlled in order to suppress potential errors.

Let us consider, in dimension 2, the example of creating a contour as a set of segments: the user provides only the points necessary to define the contour, which in turn define the lines of this contour. From the description of a line, the preprocessing consists of selecting all the points and then the set of segments that constitutes it. Lastly, it unites the segments of the different lines which form the contour. Following this simple example, we observe that:

4.6. USER DATA AND MESH GENERATOR DATA

- the effort required by the user consists only of:
  - supplying some characteristic points (cf. 4.4.1);
  - describing the useful lines (cf. 4.4.2);
  - specifying the lines constituting the contour (this list will be oriented to be able to define the *interior* of the domain described by this contour).

- the preprocessing then consists of:
  - processing the given points;
  - splitting the lines by creating intermediary points;
  - joining the contour lines while controlling the correct closure of the resulting contour;
  - checking that the data thus defined is consistent with the mesh generator envisaged.

The mesh generation package includes this preprocessing phase. The user data can be captured by the following different methods:

- creating a file containing the relevant values given in a specified format,
- creating a data file automatically:
  - by entering the relevant values manually;
  - by capturing the values interactively; in this case, the process offers graphic facilities.

Input of a geometric nature is either explicit (enumerations, equations of curves, etc.), or done by a constructive method (for example, a circle defined by three points, by tangential properties with respect to other curves, etc.).

The creation of a data file is an easy way of keeping this data in memory. Thus, at a later stage, the *history* of any value can be accessed, so that any modification can be done easily. Nevertheless, some packages do not use such a file directly: values are captured directly and the algorithms are activated immediately.

Remark 4.2 : Numerous types of data files are currently in use; among them are the following:

- files containing the requests as *key-words,* followed by a series of associated *values*; for example, a point is defined by a pre-defined list of values such as:

  - POINT

  - ...

  - 10 1.2 0.65 2.6 1

  - ...

  in this case, the meaning of the numerical values is known in advance, for example, 10 is the number of the point, 1.2, 0.65 and 2.6 are the coordinates and 1 is an attribute.

- files containing some *key-word - value(s)* couples where the type of the value is defined by the key-word preceding it, for example, for the same type of point definitions as above, we now have:

  - *POINT NUM = 10, X = 1.2, Y = 0.65, Z = 2.6, ATT = 1;* or equivalently :

  - *POINT ATT = 1, X = 1.2, NUM = 10, Z = 2.6, Y = 0.65* , the order of the values does not matter.

- files containing the definition relative to a point (using the same example), given in a constructive manner:

  - POINT 10 = (LINE a) INTER (CIRCLE c) + (constraint c_r), in this case, point 10 is the intersection of the line number a with the circle number c such that constraint c_r is satisfied (to avoid a possible ambiguity or indetermination). □

4.7 An example of bottom-up data

The bottom-up input of data is a consequence of the top-down analysis of the domain and the related physical problem.

4.7.1 Top-down analysis

Consider the example shown in figure 3.1 whose contours are shown in figure 4.3. Taking advantage of the (evident) geometrical repetitions, one can deduce that it is sufficient to consider the part of the domain shown in figure 4.4, which is split into two primal sub-sets as composed of two materials.

4.7. AN EXAMPLE OF BOTTOM-UP DATA

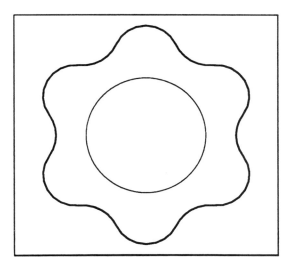

Figure 4.3: *The domain under consideration.*

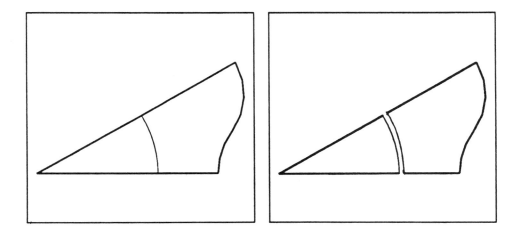

Figure 4.4: *The region to be considered and its partitioning.*

4.7.2 Bottom-up data

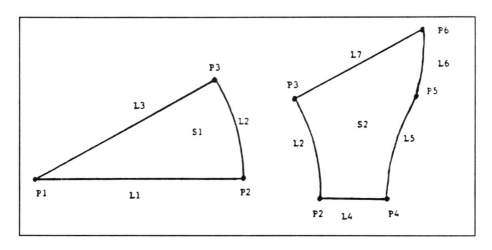

Figure 4.5: *The basic items.*

Let us assume that we use, for the mesh generation of these two subsets, methods which require the description of their contour, i.e., the data of the list of the segments which discretizes them adequately. We then define:

- the characteristic $\mathcal{P}$oints: $P1$ $P2$ ... $P6$.

  A point is described by 3 values (its attribute or reference number, and its two coordinates, x and y, cf. 4.4.1.).

Due to these points, we can create:

- the characteristic $\mathcal{L}$ines: $L1$ $L2$ ... $L7$.

  A line is defined by 6 values (its 2 ends-points, ... cf. 4.4.2.)

We are now able to describe:

- the $\mathcal{V}$olumes (in this case the $\mathcal{S}$urfaces): $S1$ and $S2$.

  A surface is defined by the oriented list of contour lines. It is therefore defined by the set of information encoded in these lines and also by the set of points of the latter. An attribute, or sub-domain number, will be assigned to each surface.

4.7. AN EXAMPLE OF BOTTOM-UP DATA

It is now possible to execute the selected mesh generator resulting in the creation of meshes $M1$ and $M2$ of surfaces $S1$ and $S2$. We then paste along line $L2$ to obtain the mesh M : $M = M1 \cup M2$ (figure 4.6 left-hand side). This mesh is then manipulated (using symmetries, rotations and pasting together) to obtain the mesh of the entire domain (figure 4.6 right-hand side).

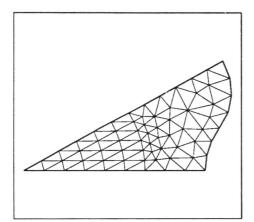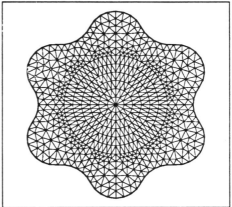

Figure 4.6: *The partial mesh and the final one.*

Notice that starting with the data of merely 6 $\mathcal{P}$oints and 7 $\mathcal{L}$ines (i.e. a little amount of information) the mesh created was simply manipulated (by symmetries, rotations and pasting) and resulted in the creation of the final mesh consisting (see figure 4.6) of 1044 elements and 559 nodes.

Remark 4.3 : The mesh shown in figure 4.6 can be compared with that of figure 5.3, which was created using a different approach. □

Chapter 5

Manual and semi-automatic methods

5.1 Introduction

This chapter presents three methods suitable for mesh creation corresponding either to a manual approach, or a semi-automatic process:

- Manual creation: all information useful for the mesh definition is supplied by the user;

- For domains with connectivity similar to a hexahedron, the mesh is derived from the definition of the distribution of points in three directions;

- More generally, for domains topologically similar to a cylinder, the volumetric mesh is derived, layer by layer, from the $2D$ mesh of any section of the domain.

The common feature in these three methods is the fact that the connectivity is either given explicitly or known implicitly. The location of the points is either input explicitly or derived from information provided by the user.

5.2 Manual method

5.2.1 Mesh generation method

This method is not based on an algorithm but consists of the enumeration, by the user, of all the elements in the mesh to be created. More precisely, the user has to provide all the information for each element relevant to the creation of the data structure associated with the mesh. To give a practical example, let us consider a mesh generator of this type, contained in the Modulef library; it is necessary to specify, for all elements:

1. its geometrical shape (segment, triangle, quadrilateral, tetrahedron, pentahedron, hexahedron or particular element);
2. its vertices assuming the topological conventions seen in section 2.3.6.;
3. the coordinates of these vertices;
4. the physical attributes of these vertices, edges and faces;
5. its sub-domain number.

Thus, the element connectivity and location of the vertices of the mesh are defined by the user.

Simple meshes, or simple sections of more complex meshes, can be created following this approach. It can also serve to create data for constructions based on modifications of simple meshes (in particular their partitioning).

5.2.2 Data description

This mesh generator needs all the information required for the creation of the relevant data structure, as data (see above).

The associated scheme is as follows

$$\mathcal{P} \to \text{Element description} \to \text{Mesh generator}$$

Reference numbers of the faces, edges and points of the created mesh are input by the user, as well as the sub-domain number assigned to each constructed element.

5.2.3 Application example in 2 dimensions

As an example (figures 5.1, 5.2 and 5.3) we show the creation of only three elements using the manual approach. This coarse mesh is then postprocessed by applying splitting, geometrical modifications and recompositions.

5.2. MANUAL METHOD

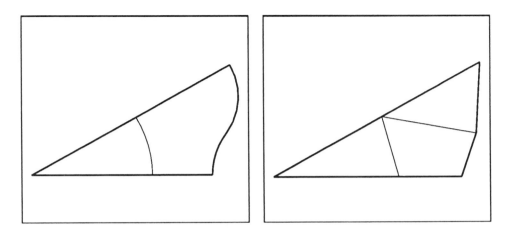

Figure 5.1: *The domain to be considered and the 3 coarse elements.*

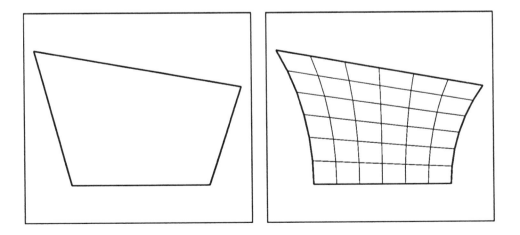

Figure 5.2: *Splitting of the quadrilateral.*

68   CHAPTER 5. MANUAL AND SEMI-AUTOMATIC METHODS

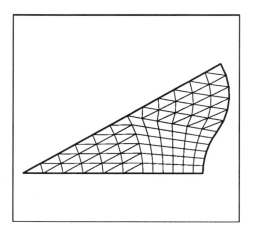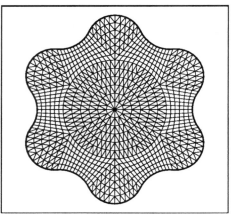

Figure 5.3: *Final mesh and mesh of the complete domain.*

After analysis of the complete domain under consideration (figure 5.3, right-hand side), only one twelfth needs to be considered (figure 5.1, left-hand side); this primal set is covered by two triangles and one quadrilateral (figure 5.1, right-hand side). The manual mesh generator constructs this very simple mesh whose elements are partitioned into N^2 elements of the same type producing a finer mesh which is closer to the geometry of the domain: some edges of the coarse mesh are defined by curves via an appropriate reference number; therefore, the points created on these edges are mapped on these curves (figure 5.2 shows the splitting of the coarse quadrilateral). The final mesh is obtained by symmetries, rotations and pasting together (see figure 5.3).

Remark 5.1 : The data consists of only 6 $\mathcal{P}$oints and 3 coarse elements. This reduced amount of values has been sufficient for the creation of the non-trivial final mesh which consists of 1296 elements and 937 nodes.   □

Remark 5.2 : It is clear that other possibilities exist to generate a mesh of this domain (see figure 4.6).   □

5.2.4 Application example in 3 dimensions

The example provided (figures 5.4 and 5.5) corresponds to the creation of only four elements by the manual approach and post-processing with splitting, geometrical transformations and recomposition, to obtain the final mesh.

5.2. MANUAL METHOD

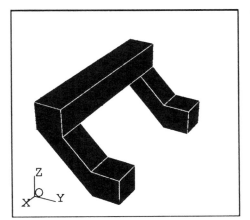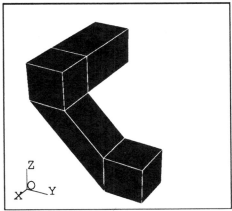

Figure 5.4: *The domain considered and its four coarse elements.*

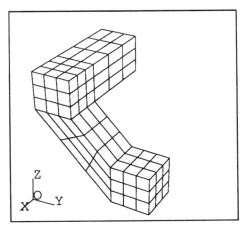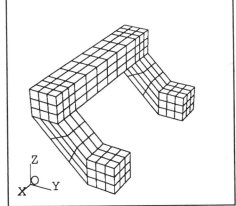

Figure 5.5: *Splitting of the primal sub-set and final mesh.*

After analysis of the complete domain (figure 5.4 left-hand side), only one half of it needs to be considered; this primal set is covered by 4 hexahedra (figure 5.4 right-hand side). The manual mesh generator produces this trivial mesh whose elements are then partitioned into N^3 elements of the same type in order to obtain a finer mesh. Finally, the desired mesh is obtained by symmetry and pasting (see figure 5.5).

Remark 5.3 : A reduced quantity of data enabled us to generate the final mesh which consists of 216 elements and 400 nodes. □

Remark 5.4 : As mentioned previously, other possibilities exist to mesh this domain. □

5.3 Hexahedral topologies

5.3.1 Mesh generation method

Suitable for cubic geometries, this method constructs the covering-up of the domain by hexahedra (which can be split later on) from the data of the points distributed in three directions.

Firstly, a mesh is created by considering only the plane defined by the first two directions. It is composed of quadrilaterals whose vertices are derived from the two sets of points corresponding to these directions. Each element of this mesh then forms a basis for the construction of the layers of hexahedra by considering the points in the third direction (figure 5.6).

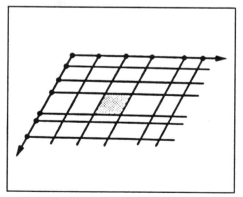 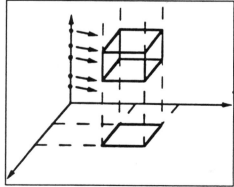

Figure 5.6: *2D partition and associated 3D construction.*

Although based on a relatively simple principle, this method solves the problem of meshes of relatively complex domains assuming this special

5.3. HEXAHEDRAL TOPOLOGIES

topology. The connectivity is of finite difference type (i,j,k) and the vertex position is defined via points given in the three descriptive directions.

This technique produces essentially hexahedral elements. Each of which can be split into two pentahedra or five or six tetrahedra (see chapter 13).

5.3.2 Data description

This mesh generator requires the values necessary for the definition of the three point distributions, as data.

The data, in this case, can be schematized as follows:

$$\mathcal{P} \to \text{Mesh generator}$$

The efficiency of the method is due to the numerous possibilities available for the definition of the vertex coordinates. As a vertex is known by its three indices, i, j and k, it suffices to provide the following data:

- Elements with constant stepsize along one, two or three directions: the value(s) corresponding to the spacing of the points in this(these) direction(s) is(are) entered;

- Elements with variable stepsizes along one or several directions: the corresponding stepsizes are input in table form;

- Arbitrary elements: the locations of the points are given in table form or as functions;

- Mixed elements: an adequate stepsize combination is given in table form or by user-functions.

The face, edge and point reference numbers of the created mesh are defined by their location (indices i, j, k), as is the case for the sub-domain numbers associated with all the created elements.

5.3.3 Application examples

We consider an example (figures 5.7) corresponding to module GEL3D1 of the Modulef library, which is based on this approach. The mesh of the left-hand side consists of 1120 elements and 840 points, that of the right-hand side consists of 180 elements and 294 points. The picture only displays the visible faces using a "shrinked" form (cf. chapter 13).

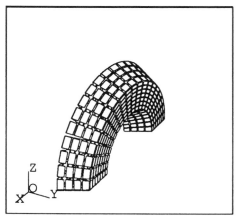 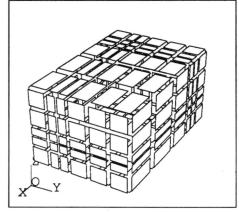

Figure 5.7: *Two examples of meshes (GEL3D1)*.

5.4 Cylindrical topologies

5.4.1 Mesh generation method

In this approach [George,Golgolab-1988], the mesh of any domain of cylindrical topology can be created from a given $2D$ mesh, the reference mesh, by stacking $3D$ elements. The nature of these $3D$ elements is derived from the corresponding $2D$ elements of the reference mesh.

In fact, any domain which it is possible to define in this manner will be seen as a domain with a cylindrical topology.

The data required by this mesh generator is therefore the $2D$ mesh of a reference section and the definition of the different sections along the generating line of the "cylinder".

Referred to as the "translation-stacking" or "extrusion" method, this approach can be seen as a generalization of the previous method.

Layers of quadrilaterals are associated with each segment of the $2D$ mesh, pentahedra are associated with each triangle, hexahedra are associated with each quadrilateral. The connectivity is derived from the two-dimensional basis mesh coupled with the index of the section under consideration.

In order to deal with domains which include the axis of the associated cylinder, it is necessary to consider cases of possible degenerations (shown in figure 5.11, right-hand side), such as a quadrilateral resulting in the

5.4. CYLINDRICAL TOPOLOGIES

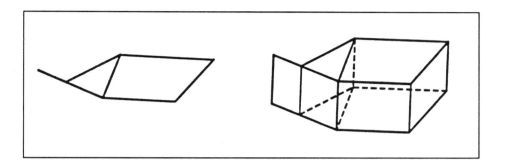

Figure 5.8: *Correspondence 2D - 3D*.

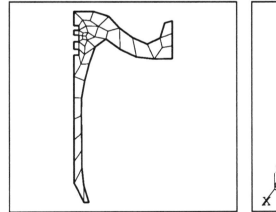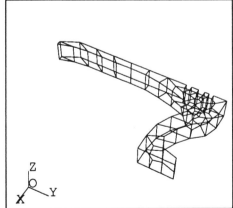

Figure 5.9: *A section and its mapping (creation of one layer)*.

creation of pentahedra instead of hexahedra.

5.4.2 Data description

As data, this method requires the two-dimensional mesh of a section, representative of the domain, and the definition of a mapping function of this section which defines the set of layers for the stacking operations.

The data can therefore be seen as:

$$2D \text{ mesh} + \text{Definition of the sections} \rightarrow \text{Mesh generator}$$

The efficiency of the method is once again due to the numerous possibilities available for the definition of the vertex coordinates of the different sections. This is done either directly from the $2D$ mesh or section by section, starting with the definition of a first section. A $3D$ vertex is known through the $2D$ vertex from which it is derived, and by its sectional index k; consequently, it suffices to provide the following data:

- Element with constant stepsize in the direction of the cylinder axis: the stepsize corresponding to the spacing of the points in this direction is specified;

- Element with variable stepsize in the direction of this axis: the corresponding stepsizes are input in table form;

- Arbitrary transport of the basis section: the transformation defining the three point coordinates as a function of the two coordinates of the corresponding basis points and the index of the section being created are entered;

- Arbitrary transport from an existing section: the transformation defining the three point coordinates of the section to be created from the three coordinates of the associated points of the existing section are specified using functions.

Both the reference numbers of the faces, edges and points, and the subdomain number of the elements created are defined by the user; these values are derived from the associated values of the generic mesh in accordance with a given correspondence.

5.4. CYLINDRICAL TOPOLOGIES

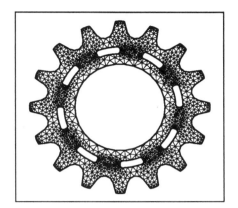
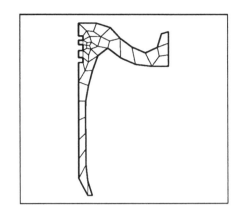

Figure 5.10: Two sets of data.

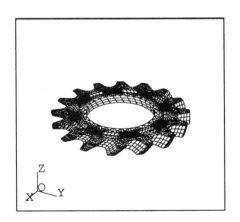
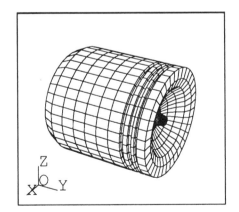

Figure 5.11: *The associated 3D meshes.*

5.4.3 Application examples

Module MA2D3E of the Modulef library corresponds to this method and meshes in figure 5.11 were obtained using this module.

The gear (figure 5.11 left hand side) corresponds to the trivial translation of the mesh shown in the left-hand side of figure 5.10, where pentahedra are associated with all triangles and hexahedra are associated with all quadrilaterals of this basis mesh. The final mesh consists of 150 pentahedra and 2090 hexahedra. The reservoir (figure 5.11 right hand side) corresponds to the rotation of the mesh of the right-hand side of figure 5.10. As the rotational axis intersects with an edge of the generic mesh, the elements derived from the quadrilaterals of this region are pentahedra. Thus the final mesh is composed of 144 pentahedra and 1224 hexahedra (in figure 5.9, the same generic mesh serves as an example for the construction of a single layer by vertical translation so that there is no problem of degeneration in this case).

The two examples (figure 5.11) show two domains considered topologically as cylinders. The gear clearly corresponds to the classical definition where the mesh results from the extrapolation of a $2D$ mesh. The reservoir can be interpreted similarly, where rotation of a $2D$ mesh was sufficient for the construction of the $3D$ mesh.

Chapter 6

Transport-mapping methods

6.1 Introduction

This chapter presents some mesh generation methods based on the mapping of a reference mesh corresponding to an elementary geometry (triangle, quadrilateral, tetrahedron, pentahedron, hexahedron, etc.); this type of method is also known as an algebraic method. The reference mesh, obtained trivially, is mapped on the real domain via a transport function. The robustness and the generality of application of the method is related to the choice of this function.

The following topics are approached:

- The general principle of the method;
- The definition of a transport function of degree 1;
- The definition of transport functions of degree 2 or higher;
- The construction of a transport function which preserves lines or faces;
- The combination of a transport function and a deformation technique.

6.2 General principle of the method

For simplicity, we consider a domain Ω with shape close to that of a quadrilateral. In this situation, the domain formally has four sides as contour and is described by the point data of this contour; we assume, in addition, that the number of points lying on two opposite sides is equal. Note that a domain will be considered as a quadrilateral if its contour can be described reasonably by four sides.

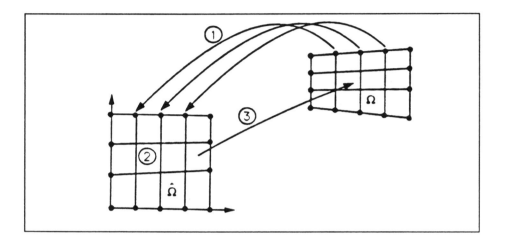

Figure 6.1: *General principle.*

Let $\hat{\Omega}$ be a unit square. We define intermediate points on the four sides of this domain, equal in number to those on the sides of the real domain, in such a way that their respective positions correspond to those of the real points.

To obtain the canonical mesh of $\hat{\Omega}$, opposite points are joined. This mesh, composed of quadrilaterals, defines the desired connectivity in a simple fashion.

Let F be a function which maps any point in $\hat{\Omega}$ to Ω while ensuring "appropriate" properties (see below), then we apply this transformation to all element vertices of the mesh of $\hat{\Omega}$. Consequently, the elements of the mesh of Ω are constructed (in terms of the point connectivity and location):

$$Basis \quad mesh \quad + \quad F \quad \Rightarrow \quad Real \quad mesh$$

The method, based on this principle, is therefore very simple. In practice, its robustness, i.e., its capacity to handle realistic geometrical situations adequately is strongly linked to the nature of function F.

We will now construct some transport functions corresponding to different mesh generation methods entering the present class.

6.3 Method based on the domain vertices

Let the real domain be analogous to a quadrilateral (considering the same example) and let a_i be the ith vertex of this quadrilateral.

6.3.1 Consistency of degree 1

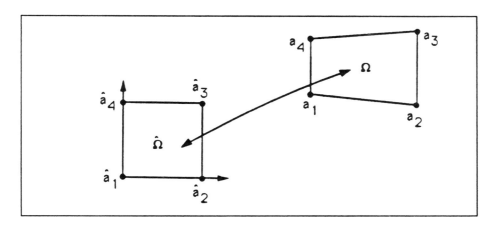

Figure 6.2: *Quadrilateral domain with straight sides.*

Function F, defined on the unit square, by:

$$F(\hat{x}, \hat{y}) = \sum_{i=1}^{4} P_i(\hat{x}, \hat{y}) a_i$$

where the P_i's are the following polynomials:

$$P_1(\hat{x}, \hat{y}) = (1 - \hat{x})(1 - \hat{y})$$

$$P_2(\hat{x}, \hat{y}) = \hat{x}(1 - \hat{y})$$

$$P_3(\hat{x}, \hat{y}) = \hat{x}\hat{y}$$

$$P_4(\hat{x}, \hat{y}) = (1 - \hat{x})\hat{y}$$

leads to the creation of image M of any vertex $\hat{M}$ of the reference -mesh in the following manner: $M = F(\hat{M})$. The mesh of Ω is obtained as the image of that of $\hat{\Omega}$ in this fashion.

As function F maps a straight line to a straight line, this process of mesh generation can only be applied in the case of a domain **with straight sides** if we want to preserve the geometry of its contour. This function defines a method called consistent to degree one.

In the case of domains analogous to elementary shapes, we choose the following expressions for function F (which are the classical interpolation functions $P1$ or $Q1$ of the finite element method):

- For a triangle : $F(\hat{x}, \hat{y}) = \sum_{i=1}^{3} P_i(\hat{x}, \hat{y}) a_i$ with:

  $P_1(\hat{x}, \hat{y}) = 1 - \hat{x} - \hat{y}$

  $P_2(\hat{x}, \hat{y}) = \hat{x}$

  $P_3(\hat{x}, \hat{y}) = \hat{y}$

- For a quadrilateral: the function given above to illustrate the method.

- For a tetrahedron: $F(\hat{x}, \hat{y}, \hat{z}) = \sum_{i=1}^{4} P_i(\hat{x}, \hat{y}, \hat{z}) a_i$ with:

  $P_1(\hat{x}, \hat{y}, \hat{z}) = 1 - \hat{x} - \hat{y} - \hat{z}$

  $P_2(\hat{x}, \hat{y}, \hat{z}) = \hat{x}$

  $P_3(\hat{x}, \hat{y}, \hat{z}) = \hat{y}$

  $P_4(\hat{x}, \hat{y}, \hat{z}) = \hat{z}$

- For a pentahedron: $F(\hat{x}, \hat{y}, \hat{z}) = \sum_{i=1}^{6} P_i(\hat{x}, \hat{y}, \hat{z}) a_i$ with:

  $P_1(\hat{x}, \hat{y}, \hat{z}) = (1 - \hat{x} - \hat{y})(1 - \hat{z})$

  $P_2(\hat{x}, \hat{y}, \hat{z}) = \hat{x}(1 - \hat{z})$

  $P_3(\hat{x}, \hat{y}, \hat{z}) = \hat{y}(1 - \hat{z})$

  $P_4(\hat{x}, \hat{y}, \hat{z}) = (1 - \hat{x} - \hat{y})\hat{z}$

  $P_5(\hat{x}, \hat{y}, \hat{z}) = \hat{x}\hat{z}$

  $P_6(\hat{x}, \hat{y}, \hat{z}) = \hat{y}\hat{z}$

- For a hexahedron: $F(\hat{x}, \hat{y}, \hat{z}) = \sum_{i=1}^{8} P_i(\hat{x}, \hat{y}, \hat{z}) a_i$ with:

  $P_1(\hat{x}, \hat{y}, \hat{z}) = (1 - \hat{x})(1 - \hat{y})(1 - \hat{z})$

  $P_2(\hat{x}, \hat{y}, \hat{z}) = \hat{x}(1 - \hat{y})(1 - \hat{z})$

  $P_3(\hat{x}, \hat{y}, \hat{z}) = \hat{x}\hat{y}(1 - \hat{z})$

  $P_4(\hat{x}, \hat{y}, \hat{z}) = (1 - \hat{x})\hat{y}(1 - \hat{z})$

  $P_5(\hat{x}, \hat{y}, \hat{z}) = (1 - \hat{x})(1 - \hat{y})\hat{z}$

  $P_6(\hat{x}, \hat{y}, \hat{z}) = \hat{x}(1 - \hat{y})\hat{z}$

  $P_7(\hat{x}, \hat{y}, \hat{z}) = \hat{x}\hat{y}\hat{z}$

  $P_8(\hat{x}, \hat{y}, \hat{z}) = (1 - \hat{x})\hat{y}\hat{z}$

6.3. METHOD BASED ON THE DOMAIN VERTICES

Remark 6.1 : This method is also applicable to the generation of meshes for triangular or quadrilateral surfaces in R^3 (points a_i are then defined by 3 coordinates).
□

6.3.2 Consistency of a higher degree

A function of the type defined in the above section does not possess sufficient properties to extend its applications to curvilinear geometries; thus, to treat more complicated cases, a different choice is required.

Considering the same example (a quadrilateral domain), we will now find a suitable function corresponding to the case of a quadrilateral with parabolic sides.

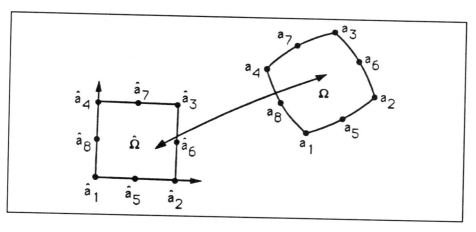

Figure 6.3: *Quadrilateral domain with parabolic sides.*

Let Ω be a domain analogous to a quadrilateral with parabolic sides, and let a_i ($i = 1, 4$) be the four vertices of this quadrilateral and a_{i+4} ($i = 1, 4$) be four points located on each of the sides, then the following function F:

$$F(\hat{x}, \hat{y}) = \sum_{i=1}^{8} P_i(\hat{x}, \hat{y}) a_i$$

with :

$$P_1(\hat{x}, \hat{y}) = (1 - \hat{x})(1 - \hat{y})(1 - 2\hat{x} - 2\hat{y})$$
$$P_2(\hat{x}, \hat{y}) = \hat{x}(1 - \hat{y})(2\hat{x} - 2\hat{y} - 1)$$
$$P_3(\hat{x}, \hat{y}) = -\hat{x}\hat{y}(3 - 2\hat{x} - 2\hat{y})$$

$$P_4(\hat{x}, \hat{y}) = (1 - \hat{x})\hat{y}(-2\hat{x} + 2\hat{y} - 1)$$

$$P_5(\hat{x}, \hat{y}) = 4\hat{x}(1 - \hat{x})(1 - \hat{y})$$

$$P_6(\hat{x}, \hat{y}) = 4\hat{x}\hat{y}(1 - \hat{y})$$

$$P_7(\hat{x}, \hat{y}) = 4\hat{x}(1 - \hat{x})\hat{y}$$

$$P_8(\hat{x}, \hat{y}) = 4(1 - \hat{x})\hat{y}(1 - \hat{y})$$

results in the construction of image M of any vertex $\hat{M}$ of the reference mesh, while ensuring that the parabolic contour contains all a_i ($i = 1, 8$). A method using this type of functions is called consistent to degree 2 (i.e. the degree of the polynomials which define it).

More generally, by increasing the degree of polynomials P_i, functions with larger application can be defined. Nevertheless, one should take care that the number of vertices defined are consistent with the geometry of the sides.

There are, for example in [Ciarlet,Raviart-1972a], polynomials leading to the construction of different functions F.

6.4 Method preserving a polygonal contour

6.4.1 Two-dimensional geometry

In this section, we will construct a function F respecting a polygonal contour exactly. Its application will clearly be more general; it caters for domains which, assuming an elementary geometrical shape (from a topological point of view), are defined by their contour and, more precisely, by the set of segments approximating it.

This approach [Gordon,Hall-1973], [Zlamal-1973], [Cook-1974] will be applicable in the case of domains topologically similar to a triangle or quadrilateral.

We suppose therefore that it is possible to define the contour of domain Ω in terms of 3 or 4 topological sides in such a way that the number of points on two logically connected sides is equal (for a quadrilateral, it means that there is the same number of points on sides 1 and 3 and sides 2 and 4; for a triangle, this constraint means that there is the same number of points on all three sides; [Burger-1976],[Vouillon] propose an heuristic method to overcome this condition). The method uses this discretization of the contour to generate the mesh by means of the three phases seen previously:

6.4. METHOD PRESERVING A POLYGONAL CONTOUR

1. The transport on the sides of the reference domain $\hat{\Omega}$ (unit triangle or square) of the points on the real contour,

2. The creation of the canonical mesh of $\hat{\Omega}$ respecting its contour points,

3. The transport of this canonical mesh onto the real domain Ω.

The elements created assume the same nature as that of the reference domain; by postprocessing, we can, in the case of quadrilaterals, subdivide them into triangles.

A detailed description of the 3 phases of the algorithm is given below:

Phase 1 : Transport of the contour points on the sides of the unit element associated with the domain. In this process, the relative distances between points are preserved (figure 6.4) in such a way that their distribution is respected. More precisely, in the case of a quadrilateral, if $(a_i, a_{i+1})_{modulo\ 4}$ is one of the sides of the real domain and b_i^k are n intermediate points located on this side, then points $\hat{b}_i^k$, located on the corresponding edge of the unit square, are created as follows:

$$\hat{b}_i^1 = \hat{a}_i \quad , \hat{b}_i^n = \hat{a}_{i+1}$$

$$\hat{b}_i^k = \frac{\sum_{j=1}^{k-1} d_{j,j+1}}{\sum_{j=1}^{n-1} d_{j,j+1}}$$

where $d_{j,j+1}$ denotes the distance between (real) points b_i^j and b_i^{j+1} (note that $\hat{b}_i^k$ represents a ratio).

Phase 2 : Canonical mesh of the unit element. In this case, this mesh (figure 6.5 left-hand side) consists of joining the points ($\hat{b}$) corresponding to one side of the unit domain to the other. In the case of a triangle, the construction is more delicate: formally the three lines, determined by the points on the edges logically linked to the desired point, are constructed; the intersection of these lines defines, in general, a triangular region (reduced to a point when the data is distributed evenly); the resulting point is then chosen inside this region (for example, the barycentre of the triangle is an immediate solution and, more generally, the point which minimizes the distance to the 3 lines). Once the canonical mesh is constructed, the connectivity of the real mesh is obtained.

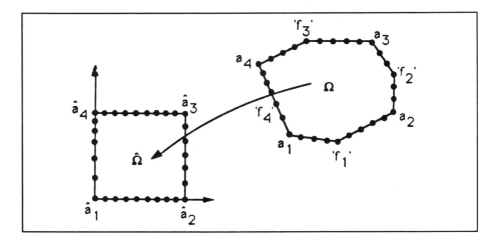

Figure 6.4: *Transport-mapping on the unit square.*

Phase 3 : Transport of this reference mesh onto the real domain. Let $\hat{M}$ be a point in the canonical mesh with coordinates $\hat{x}$ and $\hat{y}$, and let the domain be topologically analogous to a triangle, then the following transformation is used:

$$F(\hat{x},\hat{y}) = \frac{1-\hat{x}-\hat{y}}{1-\hat{x}}f_1(\hat{x}) + \frac{\hat{x}}{1-\hat{y}}f_2(\hat{y}) + \frac{\hat{y}}{\hat{x}+\hat{y}}f_3(1-\hat{x}-\hat{y})$$

$$-(\frac{\hat{y}}{\hat{x}+\hat{y}}(1-\hat{x}-\hat{y})a_1 + \frac{1-\hat{x}-\hat{y}}{1-\hat{x}}\hat{x}a_2 + \frac{\hat{x}}{1-\hat{y}}\hat{y}a_3)$$

where a_i denotes vertex i of the real domain and f_i a parametrization of side i of the same domain.

This transformation defines the desired function F which leads to the generation of M, the image of point $\hat{M}$.

In the case of a domain topologically analogous to a quadrilateral (figure 6.5), the following transformation[1] is used:

$$F(\hat{x},\hat{y}) = (1-\hat{y})f_1(\hat{x}) + \hat{x}f_2(\hat{y}) + \hat{y}f_3(\hat{x}) + (1-\hat{x})f_4(\hat{y})$$

$$-((1-\hat{x})(1-\hat{y})a_1 + \hat{x}(1-\hat{y})a_2 + \hat{x}\hat{y}a_3 + (1-\hat{x})\hat{y}a_4)$$

The f_i's are parametrizations of the sides of the domain and the a_i's the corresponding vertices: this means that the sides are approximated by the polygonal passing through points b_i^k.

---

[1]This transformation is called "transfinite interpolation", functions $x, y, (1-x)$ and $(1-y)$ are called the blending functions.

6.4. METHOD PRESERVING A POLYGONAL CONTOUR

Remark 6.2 : The above constructions use values of type $f_i(\hat{x})$ but, in practice, f_i is only known in a discrete form. It is therefore necessary to approximate the useful quantities, as near as possible, as a function of the values effectively known. □

These transformations preserve the vertices and the sides of the domain, as can be seen as an exercise (other transformations ensuring the same property can be exhibited, see for example [Perronnet-1985]).

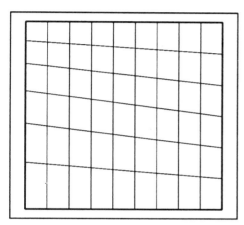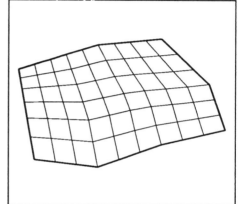

Figure 6.5: *Canonical mesh and its transport on the real domain.*

The resulting mesh is composed of elements which have the same topological nature as that of the domain under consideration.

Let n be the number of points (including endpoints) discretizing each side of a triangular domain, then the method produces a mesh consisting of $(n-1)^2$ triangles. For a quadrilateral domain, let n_1 be the number of points describing sides 1 and 3 and n_2 corresponding to sides 2 and 4, then the method creates a mesh consisting of $(n_1-1).(n_2-1)$ quadrilaterals. These elements can subsequently be split into 2 triangles, according to several options:

1. regular partitioning by specifying the diagonal (figures 6.6).

2. regular partitioning, as above, with special treatment at the "corners[2]" (figure 6.7). This method of proceeding ensures *consistency* with re-

---
[2] A corner is defined by the junction between two topological sides.

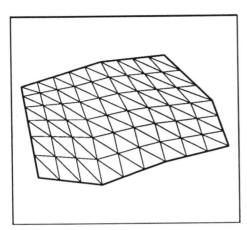 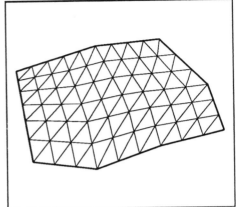

Figure 6.6: *Regular splitting.*

gard to a non-natural boundary condition (of type Dirichlet) assigned to two consecutive topological sides.

3. splitting with respect to the shortest diagonal. This operation produces the partitioning of quadrilaterals into two "optimal" triangles (figure 6.8, left hand side).

4. splitting with respect to a diagonal with special treatment at the "corners" (figure 6.8, right hand side).

Important remark : In the case of convex domains, this approach guarantees that the image of a point inside the unit reference domain is a point inside the real domain: function F is convex. This property remains valid for domains with shapes not too different from a convex; on the other hand, for overly "twisted" geometries, the image of an internal point can be evaluated outside the real domain, in which case a check is needed to validate the resulting mesh (see also section 6.5). To be rigorous, a suitable function F must satisfy the following conditions:

$$\begin{cases} F(\overset{\circ}{\hat{\Omega}}) & \subset & \overset{\circ}{\Omega} \\ F(\partial \hat{\Omega}) & = & \partial \Omega \\ \forall M \in \overset{\circ}{\Omega}, \exists! \hat{M} \in \overset{\circ}{\hat{\Omega}} & \text{such that} & F(\hat{M}) = M \end{cases} \quad (6.1)$$

6.4. METHOD PRESERVING A POLYGONAL CONTOUR 87

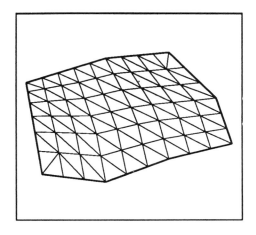

Figure 6.7: *Regular splitting with special consideration at the corners.*

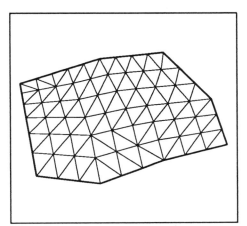
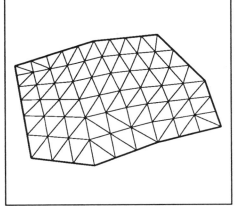

Figure 6.8: *Smaller diagonal with or without special treatment at the corners.*

It is difficult to find functions satisfying these properties in practice, particularly for non-convex domains. In addition, computed expressions are merely the discrete forms of the corresponding mathematical expressions, and consequently, the difficulty is increased. □

Remark 6.3 : With this easy-to-use mesh generator, it is not easy to obtain a variable density of elements in a given region; to do this, we either use a different approach at the mesh generation level or post-processing with local refinements (see chapter 13). □

6.4.2 Three-dimensional geometry

This type of method can be generalized to 3 dimensions, but in this case, function F is more difficult to define. In this section, we give some functions applicable to tetrahedral, pentahedral and hexahedral geometries; moreover, a slightly different method will be proposed in section 6.5.

Function F is constructed in such a way that the contour of the domain is preserved: in this case, the specified faces will be preserved.

- Case of a generalized tetrahedron:

  Let ϕ_i be any parametrization of the faces of Ω, then we consider the following function:

  $$F(\hat{x},\hat{y},\hat{z}) = \frac{\sum_{i=1}^{4} \alpha_i \phi_i(\hat{x},\hat{y},\hat{z})}{\sum_{i=1}^{4} \alpha_i}$$

  where the α_i's are functions of $\hat{x}, \hat{y}, \hat{z}$ defined as follows:

  $$\alpha_i = (1 - \hat{\lambda}_{i+3}) \frac{\hat{\lambda}_i}{(\hat{\lambda}_i + \hat{\lambda}_{i+3})} \frac{\hat{\lambda}_{i+1}}{(\hat{\lambda}_{i+1} + \hat{\lambda}_{i+3})} \frac{\hat{\lambda}_{i+2}}{(\hat{\lambda}_{i+2} + \hat{\lambda}_{i+3})}$$

  for $i = 1, 4$ with $\lambda_{i+j}, (modulo\ \ 4)$, the barycentric coordinates, i.e:
  $\hat{\lambda}_1 = 1 - \hat{x} - \hat{y} - \hat{z}$
  $\hat{\lambda}_2 = \hat{x}$
  $\hat{\lambda}_3 = \hat{y}$
  $\hat{\lambda}_4 = \hat{z}$

  therefore the restriction of F to a face of $\hat{\Omega}$ is the parametrization, ϕ_i, of this face, as can be verified as an exercise.

6.4. METHOD PRESERVING A POLYGONAL CONTOUR

- Case of a generalized pentahedron:

Similarly, the following definition provides a function with the desired properties:

$$F(\hat{x}, \hat{y}, \hat{z}) = \frac{\sum_{i=1}^{5} \alpha_i \phi_i(\hat{x}, \hat{y}, \hat{z})}{\sum_{i=1}^{5} \alpha_i}$$

where the α_i's are the functions of $\hat{x}, \hat{y}, \hat{z}$ defined by:

$$\alpha_1 = (1 - \hat{z}) \frac{1 - \hat{x} - \hat{y}}{(1 - \hat{x} - \hat{y} + \hat{z})} \frac{\hat{x}}{(\hat{x} + \hat{z})} \frac{\hat{y}}{(\hat{y} + \hat{z})}$$

$$\alpha_2 = (1 - \hat{x}) \frac{\hat{z}}{(\hat{x} + \hat{z})(1 - \hat{z} + \hat{x})} \frac{(1 - \hat{z})}{(1 - \hat{y})} \frac{(1 - \hat{x} - \hat{y})}{(\hat{x} + \hat{y})} \frac{\hat{y}}{}$$

$$\alpha_3 = (1 - \hat{y}) \frac{\hat{z}}{(\hat{y} + \hat{z})(1 + \hat{y} - \hat{z})} \frac{(1 - \hat{z})}{(\hat{x} + \hat{y})} \frac{\hat{x}}{(1 - \hat{x})} \frac{(1 - \hat{x} - \hat{y})}{(1 - \hat{x})}$$

$$\alpha_4 = \hat{z} \frac{\hat{x}}{(1 + \hat{x} - \hat{z})} \frac{\hat{y}}{(1 + \hat{y} - \hat{z})} \frac{(1 - \hat{x} - \hat{y})}{(2 - \hat{x} - \hat{y} - \hat{z})}$$

$$\alpha_5 = (\hat{x} + \hat{y}) \frac{\hat{z}}{(\hat{z} + 1 - \hat{x} - \hat{y})} \frac{1 - \hat{z}}{(2 - \hat{x} - \hat{y} - \hat{z})} \frac{\hat{x}}{(1 - \hat{y})} \frac{\hat{y}}{(1 - \hat{x})}$$

- Case of a generalized hexahedron:

For this geometry, F is given by:

$$F(\hat{x}, \hat{y}, \hat{z}) = \frac{\sum_{i=1}^{6} \alpha_i \phi_i(\hat{x}, \hat{y}, \hat{z})}{\sum_{i=1}^{6} \alpha_i}$$

where the α_i's are functions of $\hat{x}, \hat{y}, \hat{z}$ defined by:

$$\alpha_i = (1 - \hat{x}_{i-1}) \frac{\hat{x}_i}{(\hat{x}_i + \hat{x}_{i-1})} \frac{1 - \hat{x}_i}{(1 - \hat{x}_i - \hat{x}_{i-1})} \frac{\hat{x}_{i+1}}{(\hat{x}_{i+1} + \hat{x}_{i-1})} \frac{1 - \hat{x}_{i+1}}{(1 - \hat{x}_{i+1} + \hat{x}_{i-1})}$$

for $i = 1, 3$ with $\hat{x}_{i+j} = \hat{x}, \hat{y}$ or $\hat{z}$ for $j = 0, 3$ ($i + j$ modulo 4).

$$\alpha_{i+3} = \hat{x}_{i-1} \frac{\hat{x}_i}{(\hat{x}_i + 1 - \hat{x}_{i-1})} \frac{1 - \hat{x}_i}{(\hat{x}_{i-1} - \hat{x}_i)} \frac{\hat{x}_{i+1}}{(\hat{x}_{i+1} + 1 - \hat{x}_{i-1})} \frac{1 - \hat{x}_{i+1}}{(2 - \hat{x}_{i+1} - \hat{x}_{i-1})}$$

for $i = 1, 3$ with $\hat{x}_{i+j} = \hat{x}, \hat{y}$ or $\hat{z}$ as above.

The remark corresponding to the convexity of the considered domain remains valid.

In addition, other forms of function F which are applicable in the case of a generalized hexahedron can be found in [Cook-1974].

The construction of a mesh of a domain comprises the three phases described in the two-dimensional case. Phase 1, which consists of the transport of points from the real faces onto those of the unit element, follows the same principle, i.e., relative point positions are preserved. Phase 2, the construction of the canonical mesh, consists of finding these points by minimizing distances between each of them and the lines defining them. Phase 3 is identical to the two-dimensional case by choosing a function F associated with the nature of the element under consideration.

6.5 Transport and deformation

In the case of "strongly" non-convex domains, we mentioned that "internal" points can be created outside the domain. Even if this negative effect does not occur, the internal points (the image of the reference mesh points) are not always located correctly (see for example figure 6.13, right-hand side). In order to avoid this phenomenon, we propose a modification in the mesh creation process.

Four phases are now defined:

Phase 1 : The transport of the contour points onto the sides of the associated unit element while preserving the relative distances between these points, as in section 6.4.1.

Phase 2 : The creation of the canonical mesh of the unit element. This step is also identical to phase 2 in the above case.

Phase 3 : The transport of this reference mesh onto the real domain. Let $\hat{M}$ be any point in the canonical mesh, then we create the image $\tilde{M}$ of this point in the same way as in section 6.4.1.

Phase 3 is now followed by a fourth phase, point $\tilde{M}$ being only an auxiliary point in the construction:

Phase 4 : Point $\tilde{M}$ is displaced to create the desired point M, as follows:

$$\left\{ M \;=\; \tilde{M} + \frac{1}{\alpha} \sum_{P_k \in \mathcal{F}} w_k \alpha_k D(P_k) \right. \tag{6.2}$$

6.5. TRANSPORT AND DEFORMATION

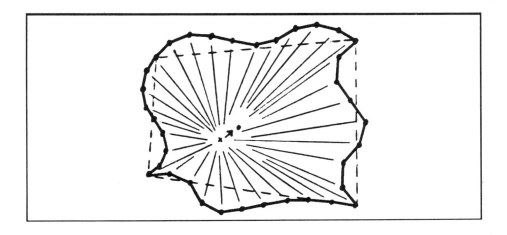

Figure 6.9: *Deformation governed by the contour points.*

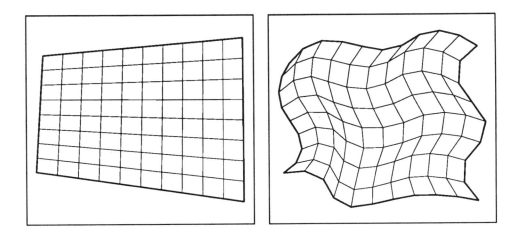

Figure 6.10: *"Straight" and final mesh.*

where we have:

$\mathcal{F}$ the set of boundary points,

$\tilde{M}$ the image of point $\hat{M}$,

w_k a weight associated with point M,

$\alpha_k = \frac{1}{d_k^\beta}$ where d_k is the distance between point $\tilde{M}$ and point P_k of set $\mathcal{F}$ and β is a parameter controlling the interaction between M and points P_k,

$\alpha = \sum_{P_k \in \mathcal{F}} w_k \alpha_k$,

$D(P_k)$ is the deformation applied to point P_k of $\mathcal{F}$: $D(P_k) = P_k - \tilde{M}$,

w_k and β are two parameters that must be adjusted.

This technique provides a better way of generating internal points, as the set of contour points is considered in a global manner (and not locally as before).

The general scheme of the mesh generation process is therefore:

$$\hat{M} \Longrightarrow \tilde{M} \Longrightarrow M$$

Remark 6.4 : The generalization of this deformation technique to the $3D$ case is obvious [Marrocco-1984]; it is used in the multiblock approach described in chapter 9. □

6.6 Data description

Let us consider the general algorithms in sections 6.4 and 6.5, which require a description of the contour of the domain to be meshed.

Concerning the data, a mesh generator based on these types of algorithms needs the input of a discretization of the contour of the domain under consideration. In dimension 2, the contour is supposed to be the union of three topological sides, in the case of a generalized triangle, and four sides, in the case of a generalized quadrilateral.

To illustrate the application of this mesh generator in dimension 2, let us consider module QUACOO of the Modulef library which corresponds to this method for a domain topologically analogous to a generalized quadrilateral (TRICOO is the module suitable for the triangular geometry). In the Modulef approach, the contour is formed by the union of characteristic

6.6. DATA DESCRIPTION

$\mathcal{L}$ines, defined by the characteristic $\mathcal{P}$oints. The data of the first line of the first side, its directional sense and the number of points on this side determine both the interior of the domain and the other sides unambiguously.

To illustrate this feature, we consider figure 6.11. To describe the domain, 9 characteristic $\mathcal{P}$oints, $P1$ $P2$ ... $P9$, are input; from these points, 9 characteristic $\mathcal{L}$ines, $L1$ $L2$ ... $L9$, are derived. Each of them is split into segments according to the definition of intermediary points.

The data of the first line of topological side 1 and of the number of points lying on this line define all the topological sides fully, as may be calculated as an exercise.

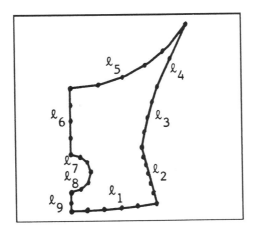

Figure 6.11: *Determination of the topological sides.*

The contour is therefore given by the following union:
$C = L1 \bigcup L2 \bigcup L3 .... \bigcup L9$. The first line of contour C is $L1$ and the number of points on side 1 is 6. It can be verified that, under these assumptions, side 4 is given by: $C4 = L6 \bigcup L7 \bigcup L8 \bigcup L9$.

Lines	1	2	3	4	5	6	7	8	9
Number of points	6	7	5	3	6	5	4	4	3

Table 6.1 : *Number of points (including endpoints) of the 9 lines.*

Remark 6.5 : A side can be formed by the union of several complete or incomplete lines (the last can only contribute by its first points). □

To illustrate the application of this mesh generator to dimension 3, let us consider module COLIBH of the Modulef library which corresponds to this method for a domain topologically analogous to a generalized hexahedron (COLIBT is the module suitable in the tetrahedral case; COLIBP corresponds to a pentahedral shape). Following the Modulef approach, the contour is formed by the set of faces of the domain, themselves defined by their edges. The latter are constructed from the characteristic $\mathcal{P}$oints.

The data of this type of mesh generator, in dimension 2, can be summarized by:

$$\mathcal{P} \to \mathcal{L} \to \text{sides} \to \text{Mesh generator}$$

and, in dimension 3, by:

$$\mathcal{P} \to \mathcal{L} \to \text{faces} \to \text{Mesh generator}$$

The reference numbers of the contour faces, edges and points of the created mesh are deduced from those of the given faces, characteristic $\mathcal{L}$ines and their items (facets, segments and intermediary points) and characteristic $\mathcal{P}$oints. Other reference numbers are set to 0. The mesh generator will assign a sub-domain number to each element of the mesh, as specified by the user.

6.7 Application examples

6.7.1 The two-dimensional case

Considering the same example (section 6.4.1), we obtain the mesh in figure 6.12. From the 9 characteristic $\mathcal{P}$oints (figure 6.11) we derived the 9 characteristic $\mathcal{L}$ines which describe the contour of the domain. From these lines the 4 topological sides are deduced after which the mesh generator creates the mesh of the domain with quadrilaterals (the option requesting the creation of triangles has not been activated).

The second example (figure 6.13) that we propose corresponds to that in figure 11.8 (where a Voronoï's method was used), and for the present approach represents a case of limited application in the sense that the domain under consideration only approximates a quadrilateral, from a topological point of view, with some imagination! Nevertheless the resulting mesh is acceptable and is composed of 414 elements and 470 points.

6.7. APPLICATION EXAMPLES

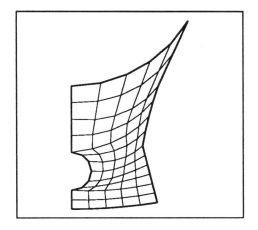

Figure 6.12: *An example of an application (QUACOO)*.

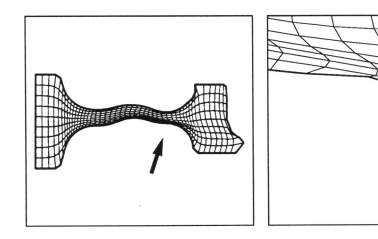

Figure 6.13: *An example of limited application (QUACOO)*.

6.7.2 The three-dimensional case

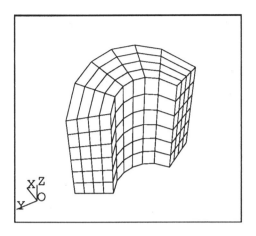

Figure 6.14: *An example of an application (COLIBH).*

This example (figure 6.14) corresponds to a domain topologically analogous to a hexahedron (it corresponds to a part of the domain shown below in figure 9.8 for the multiblock method). The 8 characteristic $\mathcal{P}$oints, the vertices of the hexahedron, are input; intermediary points are defined on the 12 edges which, on one hand, provide a more precise definition of the geometry of the real domain and, on the other hand, control the applied splitting, i.e., the point distribution and the number of subdivisions to be done and, consequently, the shape and the accuracy of the elements which constitute the resulting mesh. The final mesh contains 144 elements and 245 vertices.

Chapter 7

Mesh creation by solving a P.D.E.

7.1 Introduction

This class of methods represents an approach similar to that seen in chapter 6 with respect to the nature of domains which can be dealt with. They must be geometrically similar to quadrilaterals (in dimension 2) or hexahedra (in dimension 3). Contrary to the transport-mapping methods (cf. chapter 6), the function used to generate the mesh is not specified in advance, but it is computed by solving a predetermined **system of partial derivative equations**, chosen in an adequate way. Two possibilities exist a priori:

- Classical method: the mesh of the domain is created and the physical problem is solved on it;

- Change of variables method: a transformation function is computed which maps the real domain onto a reference domain (and conversely). The associated change of variables is affected and the solution is computed on the mesh of the reference domain.

In [Thompson-1982a], [Thompson-1985] in particular, we find a complete study of these types of methods and their different applications. This chapter only gives an outline of this approach.

7.2 General principle of the method

Let Ω be the real domain (figure 7.1, right-hand side), assumed to be topologically analogous to a generalized quadrilateral (two-dimensional case) or a generalized hexahedron (three-dimensional case). We define a reference domain $\hat{\Omega}$: a unit square or, more generally, a quadrilateral (dimension 2) and a unit cube, or, more generally, a parallelepipedon (dimension 3). This domain is meshed canonically by a structured grid (with connectivity of the type i,j or i,j,k, according to definition 1.4).

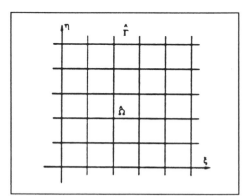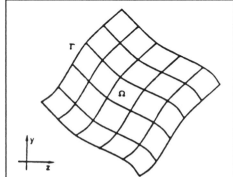

Figure 7.1: *The two systems of variables.*

Let us denote:

- x, y, z the space variables connected to the *physical* domain Ω;

- ξ, η, ζ those related to the reference domain $\hat{\Omega}$.

We then consider:

- (S) a system of partial derivative equations;

- (CL) some boundary conditions.

To each compatible couple $(S), (CL)$, for which at least one computable solution exits, a mesh generation method possessing different properties corresponds. The problem lies in the choice of this system (S), called the **generating system**. This choice must be made to ensure the creation of an acceptable and satisfying mesh possessing certain useful properties (for example, relative to the density, the orthogonality of the elements, etc.).

7.3. ELLIPTICAL METHODS

These characteristics are connected with the intended mesh usage, i.e., with the physical problem under consideration.

There are two approaches:

A- (S) is posed in terms of variables x, y, z with, as boundary conditions (CL), some isovalues specified on the four sides or six faces of the domain. (S) is then inversed, i.e., instead of computing functions $\xi(x, y, z)$, ..., we seek functions $x(\xi, \eta, \zeta)$, ... This leads to a coupled system which is solved by iterative methods (for example by relaxation), ... where the solution gives the desired result.

B- (S) is posed in terms of variables ξ, η, ζ with, as boundary conditions (CL), the exact values of the contour of Ω, i.e., the coordinates of the points on this contour. (S) is then solved in the structured grid $\hat{\Omega}$ and its solution $x(\xi, \eta, \zeta), y(\xi, \eta, \zeta), z(\xi, \eta, \zeta)$ is used for the construction of a mesh of Ω. In practice, this approach is merely a simplification of the previous analysis and produces methods which are more or less empirical with difficult theoretical proofs.

7.3 Elliptical methods

The first idea consists of using the regularizing properties of the Laplacian operator Δ, from which the method takes its name.

7.3.1 Type A approach

In this case, the two following systems are considered:

$$\begin{cases} \xi_{xx} + \xi_{yy} & = \quad 0 \text{ in } \Omega \\ \text{Boundary conditions} & \text{on} \quad \Gamma \end{cases} \quad (7.1)$$

and

$$\begin{cases} \eta_{xx} + \eta_{yy} & = \quad 0 \text{ in } \Omega \\ \text{Boundary conditions} & \text{on} \quad \Gamma \end{cases} \quad (7.2)$$

Assuming that the solution of these two systems is known, we inverse them to find $x(\xi, \eta)$ and $y(\xi, \eta)$ in order to obtain a system in terms of x and y.

Using relations:

$$dx = \frac{\partial x}{\partial \xi} d\xi + \frac{\partial x}{\partial \eta} d\eta$$

$$dy = \frac{\partial y}{\partial \xi}d\xi + \frac{\partial y}{\partial \eta}d\eta$$

and:

$$d\xi = \frac{\partial \xi}{\partial x}dx + \frac{\partial \xi}{\partial y}dy$$

$$d\eta = \frac{\partial \eta}{\partial x}dx + \frac{\partial \eta}{\partial y}dy$$

we deduce that:

$$dx = \frac{\partial x}{\partial \xi}\left[\frac{\partial \xi}{\partial x}dx + \frac{\partial \xi}{\partial y}dy\right] + \frac{\partial x}{\partial \eta}\left[\frac{\partial \eta}{\partial x}dx + \frac{\partial \eta}{\partial y}dy\right]$$

$$dy = \frac{\partial y}{\partial \xi}\left[\frac{\partial \xi}{\partial x}dx + \frac{\partial \xi}{\partial y}dy\right] + \frac{\partial y}{\partial \eta}\left[\frac{\partial \eta}{\partial x}dx + \frac{\partial \eta}{\partial y}dy\right]$$

which gives, by simple identification, the following relations:

$$x_\xi \xi_x + x_\eta \eta_x = 1$$

$$x_\xi \xi_y + x_\eta \eta_y = 0$$

$$y_\xi \xi_x + y_\eta \eta_x = 0$$

$$y_\xi \xi_y + y_\eta \eta_y = 1$$

which can be written in matrix form as follows:

$$\begin{pmatrix} x_\xi & x_\eta \\ y_\xi & y_\eta \end{pmatrix} \begin{pmatrix} \xi_x & \xi_y \\ \eta_x & \eta_y \end{pmatrix} = \begin{pmatrix} 1 & 0 \\ 0 & 1 \end{pmatrix}$$

or:

$$\begin{pmatrix} \xi_x & \xi_y \\ \eta_x & \eta_y \end{pmatrix} = \frac{1}{J}\begin{pmatrix} y_\eta & -x_\eta \\ -y_\xi & x_\xi \end{pmatrix}$$

with $J = x_\xi y_\eta - x_\eta y_\xi$, the determinant of matrix M given by:

$$M = \begin{pmatrix} x_\xi & x_\eta \\ y_\xi & y_\eta \end{pmatrix}$$

From $\xi_x = \frac{y_\eta}{J}$, we obtain ξ_{xx} in terms of the above relations:

$$\xi_{xx} = \frac{\partial}{\partial \xi}\left[\frac{y_\eta}{J}\right]\xi_x + \frac{\partial}{\partial \eta}\left[\frac{y_\eta}{J}\right]\eta_x$$

$$\xi_{xx} = \frac{\partial}{\partial \xi}\left[\frac{y_\eta}{J}\right]\frac{y_\eta}{J} + \frac{\partial}{\partial \eta}\left[\frac{y_\eta}{J}\right]\frac{y_\xi}{J}$$

7.3. ELLIPTICAL METHODS

$$\xi_{xx} = \frac{1}{J^2}[y_\eta y_{\eta\xi} - y_\xi y_{\eta\eta}] + \frac{1}{J^3}\left[-J_\xi y_\eta^2 + J_\eta y_\eta y_\xi\right]$$

The expressions for ξ_{yy}, η_{xx} and η_{yy} are formed analogously. From equation $\Delta\xi = \Delta\eta = 0$ and the values of J_ξ and J_η in terms of x and y, we deduce that x and y satisfy the following system:

$$\begin{cases} g_{11}x_{\xi\xi} + g_{22}x_{\eta\eta} + 2g_{12}x_{\xi\eta} = 0. \\ g_{11}y_{\xi\xi} + g_{22}y_{\eta\eta} + 2g_{12}y_{\xi\eta} = 0. \end{cases} \quad (7.3)$$

with:

$$g_{ij} = \sum_{m=1}^{2} A_{mi}A_{mj}$$

where $A_{mi} = (-1)^{i+m}(\text{Cofactor}_{m,i} \text{ of } M)$.

The result is a system posed in $\hat{\Omega}$ (for which a mesh exists); it is a non-linear coupled system. Relaxation techniques or, more generally, iterative methods can be used to solve this system. By imposing the real boundary conditions, this process is initialized by a given initial guess.

The analysis of the results obtained by this method, for domains with particular characteristics, and the special property requirements for the elements created, lead us to study other more complicated generation systems. In this respect, one can experiment with operators and boundary conditions.

Thus, by adding a non-zero right-hand-side to the previous system:

$$\begin{cases} \xi_{xx} + \xi_{yy} = P \\ \text{Boundary conditions} \end{cases} \quad (7.4)$$

and

$$\begin{cases} \eta_{xx} + \eta_{yy} = Q \\ \text{Boundary conditions} \end{cases} \quad (7.5)$$

we can control the distribution of points created inside the domain. The inverse system to be solved is then:

$$\begin{cases} g_{11}x_{\xi\xi} + g_{22}x_{\eta\eta} + 2g_{12}x_{\xi\eta} + J^2(Px_\xi + Qx_\eta) = 0. \\ g_{11}y_{\xi\xi} + g_{22}y_{\eta\eta} + 2g_{12}y_{\xi\eta} + J^2(Py_\xi + Qy_\eta) = 0. \end{cases} \quad (7.6)$$

using the same notation and J, the jacobian, defined by $J = det(M)$.

The role of right-hand-sides P and Q is as follows:

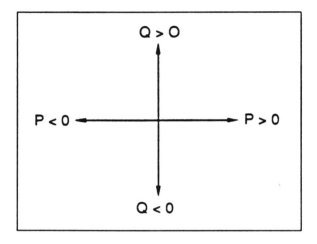

Figure 7.2: *Local effect of terms P and Q.*

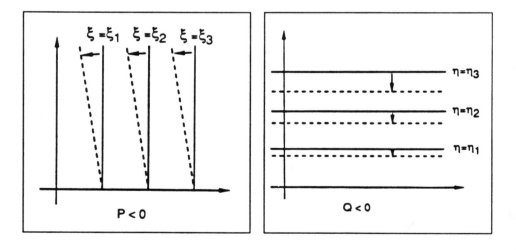

Figure 7.3: *Global effect of terms P and Q.*

7.3. ELLIPTICAL METHODS

For $P > 0$, the points are attracted to the right, for $P < 0$ the inverse effect holds (figure 7.2).

For $Q > 0$ the points are attracted to the top, for $Q < 0$ the inverse effect holds (figure 7.2).

Close to the boundary, the effect of P and Q is translated by an inclination of lines ξ or $\eta = constant$ where the points of these boundaries are prescribed. The left-hand side of figure 7.3 shows the effect of term Q on lines $\eta = constant$ and, on its right-hand side, that of term P on lines $\xi = constant$ in a region close to a fixed boundary.

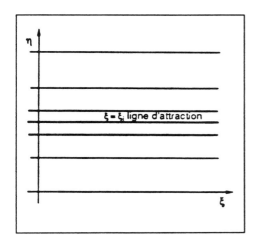
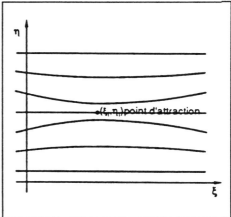

Figure 7.4: *Attraction to a line, to a point.*

P (Q) can be used to concentrate lines $\xi = cste$ or $\eta = cste$ towards a given line or to attract these towards a given point (figure 7.4). So a right-hand side of type:

$$P(\xi,\eta) = -\sum_{i=1}^{n} a_i sign(\xi-\xi_i)e^{-c_i|\xi-\xi_i|} - \sum_{i=1}^{m} b_i sign(\xi-\xi_i)e^{-d_i[(\xi-\xi_i)^2+(\eta-\eta_i)^2]^{\frac{1}{2}}}$$

where n and m denotes the number of lines in ξ and η of the grid respectively, produces the following effect:

for $a_i > 0$, $i = 1, n$, lines ξ are attracted to line ξ_i,

for $b_i > 0$, $i = 1, m$, lines ξ are attracted to point (ξ_i, η_i).

These attractions are modulated by the values of a_i (b_i) and by the distance to the attraction line (attraction point), this distance is itself modulated by coefficients c_i and d_i.

For $a_i < 0$ $(b_i < 0)$, the attraction is transformed into a repulsion. For $a_i = 0$ $(b_i = 0)$, there is no particular action connected to line ξ_i (or to point (ξ_i, η_i)).

A right-hand-side Q of the same form produces analogous effects with respect to η by interchanging the roles of ξ and η.

Refer to [Thompson-1985] where other forms of right-hand sides P and Q producing other properties are discussed (for example, the concentration of lines ξ or η towards an arbitrary line and not only towards a particular one ($\xi = cste$ ou $\eta = cste$) or towards a given point to increase the mesh density near this point.

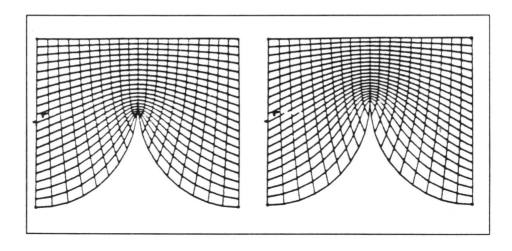

Figure 7.5: *Initial and corrected meshes (by Doursat).*

Refer to [Doursat,Perronnet-1989] for an investigation of a method of this type based on the [Winslow-1967] approach coupled with a slightly different control for the improvement of the element shapes, in particular close to the boundary (in the case of a domain with strongly non-convex regions). Figure 7.5 shows a possible application of this method.

Moreover, to obtain orthogonality properties, a generation system of the following form could be chosen:

7.3. ELLIPTICAL METHODS

$$\begin{cases} g_{11}x_{\xi\xi} + g_{22}x_{\eta\eta} + g_{11}g_{22}(x_\xi\Delta\xi + x_\eta\Delta\eta) &= 0. \\ g_{11}y_{\xi\xi} + g_{22}y_{\eta\eta} + g_{11}g_{22}(y_\xi\Delta\xi + y_\eta\Delta\eta) &= 0 \end{cases} \quad (7.7)$$

7.3.2 Type B approach

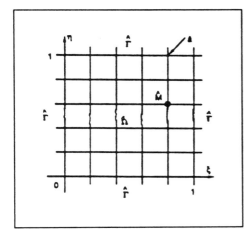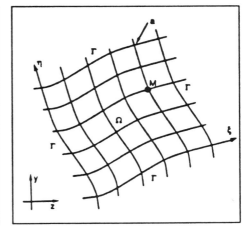

Figure 7.6: Transformation of grids.

Let $\hat{\Omega}$ be a unit square (to facilitate this discussion, the two-dimensional case is assumed) and let a regular grid cover this domain (figure 7.6, left-hand side). Let $\hat{\Gamma}$ denote the boundary of $\hat{\Omega}$, so that:

$$\hat{\Gamma} = \left\{(\eta,\xi), (\eta = 0) \bigcup (\xi = 1) \bigcup (\eta = 1) \bigcup (\xi = 0)\right\}$$

then we consider the following two systems:

$$\begin{cases} x_{\xi\xi} + x_{\eta\eta} &= 0 \text{ in } \hat{\Omega} \\ x(\xi,\eta) &= x_\Gamma \text{ on } \hat{\Gamma} \end{cases} \quad (7.8)$$

and

$$\begin{cases} y_{\xi\xi} + y_{\eta\eta} &= 0 \text{ in } \hat{\Omega} \\ y(\xi,\eta) &= y_\Gamma \text{ on } \hat{\Gamma} \end{cases} \quad (7.9)$$

The boundary conditions consist of imposing the real value at any point $\hat{a}$ of $\hat{\Gamma}$, namely:

$$x(\xi_{\hat{a}}, \eta_{\hat{a}}) = x_a$$

$$y(\xi_{\hat{a}}, \eta_{\hat{a}}) = y_a$$

The solution of these two generation systems gives the values $x(\xi, \eta)$ and $y(\xi, \eta)$ at each point $\hat{M}$ of $\hat{\Omega}$.

Consequently, the mesh of Ω is known: its connectivity is that of the canonical mesh of the reference domain $\hat{\Omega}$, the vertex positions are determined by the couples (x, y).

One interpretation is that the elements sought are deduced from the solution x, y of two sets of equipotential lines.

All methods of this type are in fact derived from the above analysis in approach [A] by neglecting certain relationships between variables. In this way, the difficulties related to the non-linearity and the coupling terms disappear.

In this type of mesh generation, the elements created are not guaranteed valid. It is, in practice, possible to create points outside domain Ω. This negative effect occurs in the case where the boundary of Ω contains pronounced concave sections. Consequently, the generation system is modified by introducing adequate corrective terms to avoid these types of negative effects.

7.4 Other methods

Based on the same principle (type [A] approach), generation systems based on operators other than the Laplacian can be written. Hyperbolic and parabolic methods enter into this class of methods. Therefore:

$$\begin{cases} x_\xi x_\eta + y_\xi y_\eta = 0 \\ x_\xi y_\xi - x_\eta y_\xi = V(\xi, \eta) \end{cases} \quad (7.10)$$

is a hyperbolic generation system controlled by the given function $V(\xi, \eta)$, which operates on the surface of the elements created. This approach produces, by construction, an orthogonal mesh, whereas in constrast the control term $V(\xi, \eta)$, used as data, is not always simple to evaluate. Moreover, the exterior boundary cannot be defined explicitly[1] and, consequently, the method only results in the mesh of the exteriors of the domains without precision on their limits. Figure 7.7 shows (left-hand side) the application of this method to the construction of a mesh of the exterior of a shuttle nose. The right-hand side presents the mesh resulting from the splitting of the preceding mesh [Désidéri-1990].

---

[1] without additional effort.

7.4. OTHER METHODS

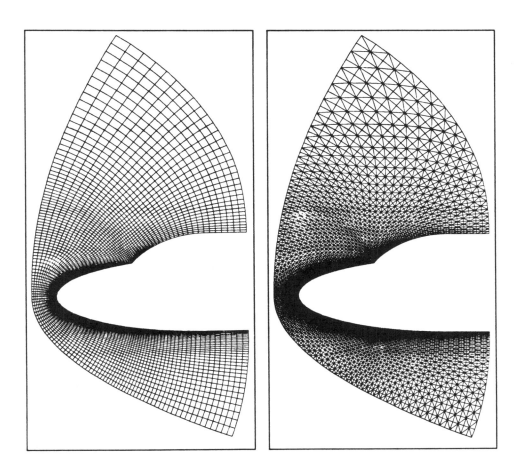

Figure 7.7: *Meshes constructed by a hyperbolic method (by Désidéri).*

7.5 Some remarks

1. For arbitrary geometries it is necessary to either define an artificial partitioning for the expression of the contour in terms of a contour similar to that of a quadrilateral (hexahedron) (see figures 7.8, 7.9 and 7.10) or decompose the domain into several blocks having the desired topology (cf. chapter 9), in these types of methods.

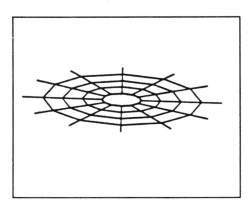
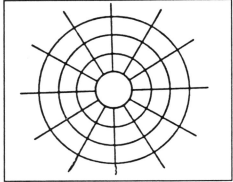

Figure 7.8: *O-type decomposition.*

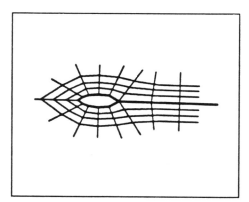
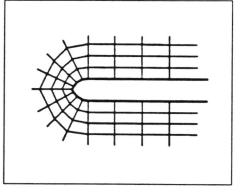

Figure 7.9: *C-type decomposition.*

In order to find the quadrilateral domain assimilable to the real domain, three types of geometries are usually distinguished (and three types of induced meshes). According to the shape of the domain and

7.5. SOME REMARKS

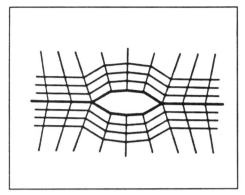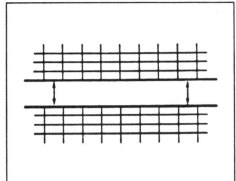

Figure 7.10: *H-type decomposition.*

to the type of result sought, this similarity will be of *O-type*, *C-type* or *H-type* (see figures 7.8, 7.9 and 7.10). For the cases where this analogy is not possible, we define artificial cuts in order to attain the known situation.

2 The application to the mesh generation of surfaces in R^3 is one of the possible extensions of these methods.

3 A popular class of applications of this approach is the space discretization of fluid mechanics problems in which we seek to obtain orthogonal elements and to enjoy the structured (i, j, k) connectivity associated with this kind of mesh generator.

Two examples of exterior meshes of a profile are found in [Joly-1988]. The mesh of the right-hand side was produced by this type of method while that on the left-hand side was created using an advancing front method (cf. remark 10.8). The main interest of the P.D.E. approach, leading to the direct creation of the i, j connectivity in the created meshes, is clearly visible in these applications, whereas the same result is not so easily attainable for other approaches (cf. same remark).

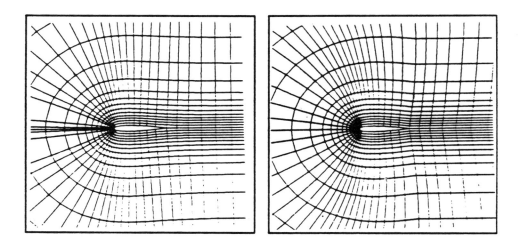

Figure 7.11: *Two examples of meshes around a profile (by Joly).*

Chapter 8

Grid superposition-deformation

8.1 Introduction

This type of generator was investigated by [Yerri,Shephard-1984], [Cheng et all-1988] and [Shephard et al. 1988]. This method constructs a mesh of the domain Ω under consideration, essentially from the data of points on its contour. A regular grid, or a grid based on a quadtree construction in 2 dimensions, or on an octree construction in 3 dimensions, is defined in such a way as to contain domain Ω. It is composed of square or cubic boxes whose size is a function of the data (the contour). This partitioning may then be deformed to resemble the real geometry of Ω more accurately, after which these boxes can be split into triangles or tetrahedra (in 3 dimensions), following a smothing phase, in order to produce the desired mesh.

The following items are discussed in this chapter:

- The presentation of a method using a regular grid;

- The presentation of a method based on a tree type grid. Three aspects are investigated:

    - the definition (within our framework) of the notion of a tree associated with a grid;
    - the construction of a method using a quadtree in the two-dimensional case;
    - the extension of this method to the three-dimensional case.

8.2 Generation using a regular grid

The domain Ω is merged into a regular grid G composed of squares (dimension 2) or cubes (dimension 3) of identical size. This size is a function of the smallest distance between two contour points of the domain, and depends also on the available resources. The process comprises the following phases:

- removal of boxes of G which do not intersect the domain Ω;

- processing of boxes of G containing a section of the boundary of the domain; two variations are generally employed:

    - a box whose intersection with the boundary is not empty is considered as an element of the mesh; in this way, the final covering-up will only be an approximation of the given domain (the accuracy of this approximation depends on the stepsize of the boxes of the chosen grid),

    - this type of box is modified in such a way that the boundary is better approximated.

- enumeration of the mesh elements:

    - a purely internal box becomes an element of the mesh;

    - a box intersecting the boundary can be considered by itself as an element of the final mesh, in which case (figure 8.2, left-hand side) the result approximates the domain as well as possible, or can be split into one or several triangular or quadrilateral elements (figure 8.2, right-hand side) (dimension 2), or tetrahedral or hexahedral elements (dimension 3). The splitting of such a box relies on the analysis of the different configurations possible (see section 8.4 in which some suitable solutions for this splitting are listed).

Simple in principle, this method is not easily adaptable to the case where the discretization of the contour, serving as data, is irregular: this situation generally requires a very fine stepsize and then may lead to an overflow in the number of boxes for grid G, and consequently in the number of elements of the resulting mesh. Consequently, the alternative approaches proposed in the following sections are of interest.

On the other hand, for certain specific applications (for example for forecast computation), a relatively regular mesh will be a source of simplification at the computational level.

8.2. GENERATION USING A REGULAR GRID

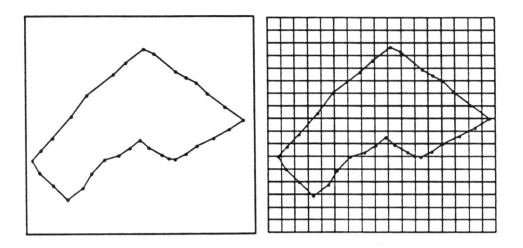

Figure 8.1: *The contour of the domain and the initial grid.*

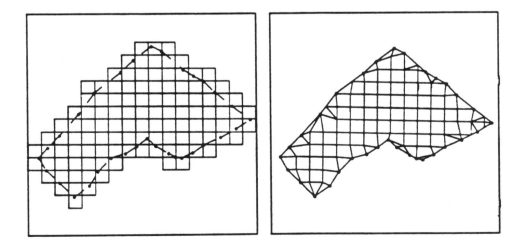

Figure 8.2: *Removal of the exterior and final meshes (case 1 and 2).*

8.3 Quadtree and octree

In this short section, we recall the definition of a quaternary or octal tree. The presentation of this technique is done in the present framework (mesh generation), to determine the use of this structure to construct a grid G enclosing domain Ω.

Considering the two-dimensional case, a square (or quadrilateral) is defined in such a way that Ω is entirely covered by this element; let G be this square, and c_1 its only box.

Let C_r be a criterion for the splitting of any box in G, then we define the splitting as follows:

c_i, the **parent**, is split into four boxes, or **daughters**, denoted c_{ij} for $j = 1, 4$ via the creation of the medians with respect to the edges of c_i.

Starting from $G = c_1$, we create, step by step, the grid $G = c_{ijkl...}$ for $i = 1, 4$, $j = 1, 4$, ..., using a recursive subdividing of boxes c of G until the criterion C_r is satisfied.

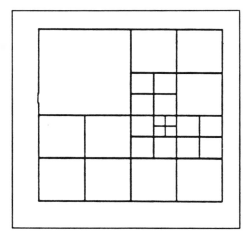
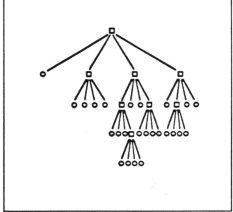

Figure 8.3: *Quadtree data structure.*

This technique gives, for every element $c_{indices}$, the exact knowledge of its rank in the hierarchy of the splitting thanks to the index $_{indices}$ encoding the position of the box c under consideration. Consequently, the position of any box is easily known.

It is obvious that the structure constructed is a *quadtree* in the two-dimensional case.

In 3 dimensions, the method starts with a cube (or a quadrilateral parallelepipedon) which is recursively split, according to criterion C_r, into 0 or 8 daughters, to obtain a representation with an *octree structure*. As a consequence, the partitioning G of the space serves as a control (one or several items of information are associated with each existing box, which are easily accessible thanks to the present tree structure).

It remains to specify criterion C_r precisely. We refer the reader to sections 8.4 and 8.5 which use this structure and suggest an example of this type of criterion adapted to our mesh generation problem. With regard to this kind of representation, based on a tree structure, the reader is also referred to [Meagher-1982], [Aho et al. 1983] and [Samet-1984].

8.4 Generation based on a quadtree

Let Ω be a domain in R^2 described via a polygonal discretization of its contour. We will create a grid G assuming the quadtree structure as mentioned above, after which this grid will be used to obtain a mesh of Ω. Based on the notion of a quadtree, this method is called the **"quadtree"** method. It consists of the following phases:

- Phase 1 : A first quadrilateral is formed which contains all the contour points. The method consists of splitting, starting with this initial box, each quadrilateral (or parent) into 4 quadrilaterals (or daughters), using a recursive process, until the final partitioning of the initial quadrilateral into quadrilateral elements is obtained where each of these elements contains at most one contour point, this is the C_r criterion. The property defining C_r serves as a control of the size of each element (or box or quad) of G (figure 8.5, left-hand side).

- Phase 2 : The partitioning resulting from the above step can be balanced (figure 8.5, right-hand side). This process consists of continuing the splitting, by defining daughters, until at most only one intermediary point exists on an edge of a box.

- Phase 3 : An analysis of thus created elements is performed. The possible configurations or *patterns* are underlined and then treated. The choice of C_r (see below) induces the following patterns:
    - external quad:
        * this type of quad does not interest us and will be eliminated.
    - internal quad (figure 8.6):

CHAPTER 8. GRID SUPERPOSITION-DEFORMATION

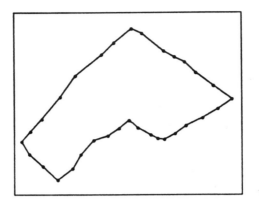

Figure 8.4: *The contour of the domain.*

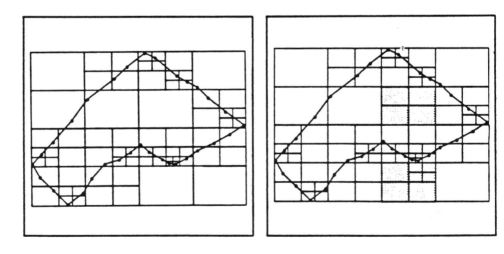

Figure 8.5: *Initial structure and balancing of the grid.*

8.4. GENERATION BASED ON A QUADTREE

* without any point on its sides (pattern 0),
* with one point on one side (pattern 1),
* with a point on two consecutive sides (pattern 2),
* with a point on two non-consecutive sides (pattern 3),
* with a point on three sides (pattern 4),
* with a point on each of its sides (pattern 5).

– quad which intersects the contour (figure 8.7):

* the contour point included in the box is close to a vertex of the box (pattern α),
* the contour point is "clearly" internal with respect to the box; in this case, the points, intersections of the boundary with the sides of the box, are created and then the proposed splitting of the box depends on the presence or not of some grid points on the sides of the box (patterns β, γ, etc).
The distinction between the different cases is made to avoid degenerate situations.
We notice that the treatment of the boxes close to the boundary is one of the difficulties of the method.

A different criterion C_r clearly implies different patterns. The situations and the patterns having been identified, the treatment of the grid boxes is as follows:

– External quad not including a point: such a quad is removed,

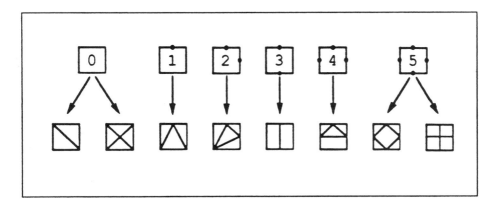

Figure 8.6: *The different patterns and their treatment (internal part).*

– Internal quad not including any contour point: such a quad produces a quadrilateral (which can be split into triangles) if its sides do not include any intermediary points (pattern 0), or is split into triangles, or possibly into quadrilaterals for the inverse case (patterns 1, 2, 3, 4, 5),

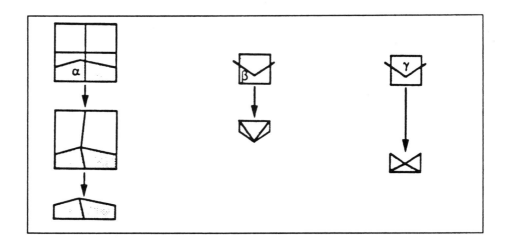

Figure 8.7: *The different patterns and their treatment (the contour).*

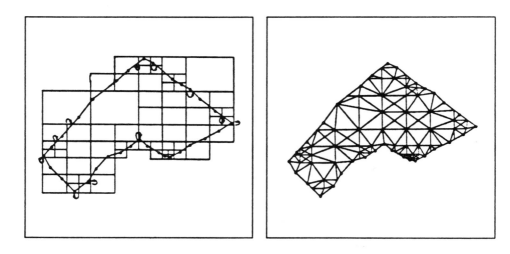

Figure 8.8: *Removal of the exterior and resulting mesh (before smoothing).*

8.5. GENERATION BASED ON AN OCTREE

– Quad including a "piece" of the contour: The intersection points of the contour and sides of the quad are created; we define, in this way, a partitioning of the quad where only the part internal to the domain is retained (*the final contour* of the mesh is thus created at this stage). A variation, using a slight deformation of the boxes of the grid G, leads to the identification of a real point with a grid point, or a point resulting from the intersection of the contour and the grid, for the case where such points are close.

The creation of the elements in the final mesh is done as a consequence of the enumeration of the different patterns possible. One or several splitting possibilities for the creation of a conformal mesh corresponds to each pattern, the solution of which depends on the identified pattern and on its neighbours. The existing variations of this mesh generation method (remark 8.2) concern the choice selected for the definition of this splitting (for a fixed criterion C_r).

- Phase 4 : A regularization of internal points is then effected (the internal points are the vertices (excluding those on the contour) of the quad of the partition resulting from step 2 of this process). This phase, usually based on a barycentrage of the vertices with respect to their neighbours, provides a better balanced result; to avoid very flat elements, a technique of diagonal swapping can be applied (such a process, run iteratively, is applied to each pair of triangles sharing a common edge).

Remark 8.1 : The initial contour is not, in general, present in the final mesh. The latter respects the initial geometry but may contain a slightly different boundary discretization (a given edge being the union of smaller edges). □

Remark 8.2 : There are numerous variations for this method. As mentioned above (phase 3 of the process), they correspond to different treatments of patterns. In addition, they can also be linked to different enumerations of known patterns. □

8.5 Generation based on an octree

The method presented in the two-dimensional case can be extended to 3 dimensions [Yerri,Shephard-1984],[Shephard et al. 1988], while the notion of a quadtree is replaced by that of an octree: a parent is an hexahedron which is used to define 0 or 8 daughters in accordance with the splitting criterion C_r. The splitting of the partitioning composed of hexahedra is therefore a more complex process than in dimension 2. In particular, to obtain the conformity of the resulting mesh, all elements are split into tetrahedra.

The phases of the algorithm are of the same nature as for dimension 2:

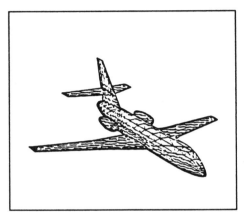

Figure 8.9: *Octree associated with a complete aircraft (doc. Golgolab).*

- Phase 1 : An initial quadrilateral parallelepipedon is formed which contains all the boundary points. The method consists of splitting, from this initial grid, each box (or parent) into 0 or 8 boxes (or daughters) by a recursive process until the final partitioning is obtained in which case the boxes contain at most one contour point (criterion C_r). This property enables the control of the size of each element in G. Figure 8.9 shows the boxes of the octree associated with a complete aircraft; only the boxes with a non-empty intersection with the domain are shown (in this figure, it can be seen that the octree is not symmetric although the domain is symmetric, due to the fact that the centre of gravity of the octree does not coincide exactly with that of the aircraft).

- Phase 2 : The partitioning resulting from the above step can be smoothed by the definition of daughters in such a way that at most one intermediary point exists on one side (therefore on a face) of a box.

8.5. GENERATION BASED ON AN OCTREE

- Phase 3 : An analysis of quads created thus is made:

  - An external quad containing no points is removed,
  - Every internal quad containing no point produces a hexahedron which is then split into tetrahedra in order to obtain a conformal result,
  - For every quad containing a "piece" of contour, the points of intersection between the contour and the edges of the quad are computed, after which a partitioning of the quad is defined of which only the internal part is retained (*the final contour* of the mesh is created at this stage). A variation by slight deformation of the boxes of grid G allows the identification of a real point with a point resulting from the intersection of the contour with the grid, for the case where these two points are close.

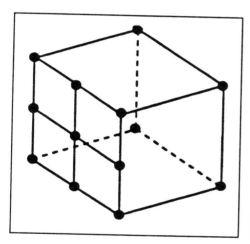 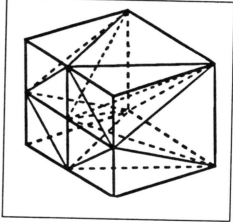

Figure 8.10: *Creation of tetrahedral elements (analysis of the patterns).*

The full enumeration of the different *patterns* possible leads to a larger number of cases than in dimension 2 and, in addition, the splitting possibilities of the patterns resulting in conformal elements are more delicate to define. In particular, the creation of tetrahedra can lead to the removal or modification of some vertices, edges or faces of the initial grid (excluding the contour). The creation of tetrahedra is done in two steps [Grice et al. 1988]:

- the meshing, in triangles, of the faces of the pattern selected to be compatible with the faces of the neighbouring patterns,
- the incorporation of these triangles to create the tetrahedral elements.

- Phase 4 : A regularization of the internal points is then performed as in 2 dimensions.

The meshes resulting from this "quadtree" or "octree" approach are generally of good quality in the regions relatively far away from the boundaries; however, the quality may be poor in their vicinity. The quality of the elements created depends on the list of defined patterns and on their treatment. In addition, one must notice that the accuracy of the boundary discretization can induce a very large number of elements.

To conclude, we note that the "quadtree" or "octree" technique can be applied advantageously as a support for advancing front or Voronoï mesh generation methods (cf. chapters 10 and 11). The knowledge of the boxes of the grid is, in these cases, used to determine the position of a point, edge or face in relation to their surroundings.

Chapter 9

Multiblock methods

9.1 Introduction

Multiblock methods, sometimes referred to as super-element methods, have been introduced by numerous authors, for example [Thompson-1987], [Fritz-1987], [Pierrot et al. 1979] and [Ecer et al. 1985]. The main ideas of these methods are discussed in this chapter, one of them ([George-1989d]) being fully detailed.

The multiblock approach, especially in the three-dimensional case, represents a solution for the creation of the mesh of geometrically complex domains, and for which the methods described previously are either ill adapted or fail. Another promising application of the multiblock approach is that it is well suited for parallel computation.

The basic idea consists of partitioning the domain into a series of blocks of elementary topology (triangular, quadrilateral, tetrahedral, pentahedral and hexahedral). Each block of this coarse partitioning is then processed by one of the methods described in chapters 6 and 7. The mesh of the entire domain is obtained by pasting together the different meshes of the blocks of the initial partitioning.

The multiblock technique induces two kinds of difficulties concerning:

- the creation of adequate elementary blocks and their meshes; the nature of the final mesh depends on the choice of these blocks;

- the management of the block interfaces to ensure the correct pasting together of their meshes[1].

---
[1] It is noticed that some multiblock methods, namely the hybrid multiblock methods, do not need the conformal requirement at the block interfaces.

9.2 Multiblock methods in 2 dimensions

9.2.1 Mesh generation algorithm

Adapted to an arbitrary geometry, this type of mesh generator creates the covering-up of the domain under consideration by segments, triangles and quadrilaterals, starting from a coarse mesh of this domain composed of the same elements and control points defined on the edges of the coarse elements or blocks. These points are used to split each block into sub-elements of the same type.

The method consists of four steps:

Step 1 : The construction of a coarse partitioning of the domain and the definition of the control points (figure 9.2, left-hand side), i.e. :

- the definition of the vertices of the blocks;
- the enumeration of each block in terms of vertices;
- the definition of the *control points* on the edges of the blocks: number of points and their position; in order to re-encounter the case described in chapter 6, the same number of control points on two logically connected edges will be assumed.

Step 2 : The splitting of each block in accordance with the position and the number of points on its edges; this process consists of:

- the mapping onto the reference element, of the same geometrical nature as the block considered, of the control points;
- the canonical splitting of this reference element with respect to these points;
- the transport of the preceding mesh onto the real block (cf. 6.4.1.).

This operation uses the transport functions of the type defined in chapters 6 or 7. The only difficulty lies in the correct enumeration of the element vertices, i.e., in the construction of the connectivity (this is the purpose of step 3).

Step 3 : The enumeration of the elements resulting from the splitting: to obtain the elements composing the mesh of the block, it suffices to enumerate their vertices. To do this, a formal numbering, in terms of i and j, is used after which the global numbers of the vertices are formed via a pointer $P(i,j)$, constructed for this purpose.

9.2. MULTIBLOCK METHODS IN 2 DIMENSIONS

To detail this process, let us consider the case of a block with quadrilateral shape. The couples $(0,0), (n_1, 0), (n_1, n_2)$ and $(0, n_2)$ denote the formal numbers of the four coarse vertices of the block (n_1 and n_2 represent the number of points - 1 (or the number of subdivisions) on edges 1 and 3, and edges 2 and 4 of this block respectively):

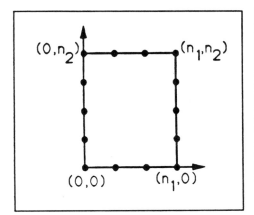
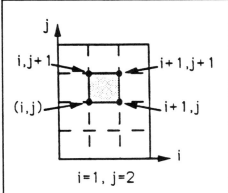

Figure 9.1: *Formal numbering.*

The indices, i and j, of any vertex referenced in the space defined by the four vertices, allow us to associate the couple (i, j) with this vertex.

$$i, j \Rightarrow (i, j)$$

It is clear that any element of the splitting is known (figure 9.1, right-hand side) by its four vertices corresponding to the couples $(i, j), (i+1, j), (i+1, j+1)$ and $(i, j+1)$. This formal numbering makes the computation of the global numbering of the vertices possible:

$$(i, j) \Rightarrow N = P(i, j)$$

where pointer P is created as follows:

- if vertex S is a vertex of the coarse mesh, $P(i, j)$ is its number in this mesh;
- if vertex S is located on a coarse edge, $P(i, j)$ is calculated thanks to the number of this edge in the coarse mesh (see section 13.9);

- if S is an internal point, $P(i,j)$ is defined sequentially by the first free number (i.e. after the last number of the last point of the last coarse edge).

Knowing P, it is easy to find the four vertices of any element from the four couples $(i,j),(i+1,j),(i+1,j+1)$ and $(i,j+1)$ associated with the i,j numbering: they have the following numbers: $P(i,j), P(i+1,j), P(i+1,j+1)$ and, finally, $P(i,j+1)$.

Step 4 : The pasting together of the meshes of each block to obtain the final mesh (figure 9.2, right-hand side): the interfaces between these blocks are the partitioning of the coarse edges; consequently, the pasting process is implicit for two adjacent blocks since the numbering of the vertices of the partitioning of the edges is global.

9.2.2 Data description

As data, this mesh generator requires the input of a coarse discretization of the domain in terms of blocks of linear, triangular and quadrilateral nature and, in addition, a distribution of points on the edges of these blocks. The number of these points must be consistent, so that two logically connected edges must be discretized with the same number of control points.

In the case of "straight" edges, the control points can be generated automatically and it suffices to specify their number and desired distribution (for a given ratio); for "curved" edges, the control points are given explicitly in order to describe the shape of the "curve" precisely.

The data of this mesh generator is therefore of the form:

$$\text{Blocks + Control points} \rightarrow \text{Mesh generator}$$

The edge and point references of the created mesh are derived from those of the coarse mesh, while the references of the items not processed in this way are set to 0. The mesh generator associates a sub-domain number, provided by the user, to all elements of the mesh belonging to the same block.

9.2.3 Application example

Module COLIB2 (option NDIM = 2, this mesh generator also applies to dimension 3) of the Modulef library corresponds to a method based, for

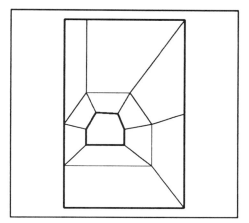 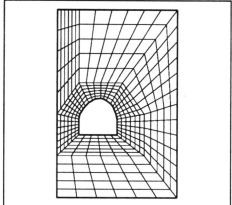

Figure 9.2: *Coarse mesh (data) and final mesh (COLIB2).*

each block of the coarse mesh, on the approach described in chapter 6. The following application example was produced by this module: the resulting mesh (figure 9.2) consists of 366 elements and 438 nodes, the coarse mesh (i.e. the set of blocks) only consists of 11 elements, of quadrilateral type, which have been created from only 18 points.

9.3 Multiblock methods in 3 dimensions

9.3.1 Mesh generation algorithm

For an arbitrary geometry, this approach [George-1989d] results in the construction of the desired covering-up starting from the structured partitioning of a coarse mesh of the domain. This partitioning is composed of blocks of simple geometry (line, triangle, quadrilateral, tetrahedron, pentahedron and hexahedron). The splitting is governed by the contour points (on faces and edges) of the coarse mesh and is dependent on their relative positions.

In principle, this method is identical to that described in the two-dimensional case (cf. the above section) and can be summarized as follows:

> Step 1 : The incorporation or the creation of the mesh of the faces of the different blocks. In this respect, we distinguish two cases:
>
> - The mesh of the faces is given: we assume that it corresponds to a "structured" mesh. Each triangular face is split into N^2

sub-triangles by the canonical method, each quadrilateral face is split into N^2 sub-quadrilaterals in the same way;

- The mesh of the faces is created from the points on their edges: the process seen in the 2D situation is used (cf. section 9.2.1) to generate the partitioning of each face into sub-elements.

Step 2 : The creation of the mesh of the blocks. Considering each element of the coarse mesh and the mesh of their faces, the mesh of a block is created by partitioning. There are two possibilities:

- using one of the methods seen in chapter 7;
- using one of the methods described in chapter 6.

A variant of the latter is now presented consisting of:

- the transport onto the faces of the unit element with the same nature as the block, of the points of the faces of this unit element, by conserving the relative distances between these points;
- the canonical mesh of the unit element. This mesh is produced by connecting the points corresponding logically from one face of the unit domain to another (cf. 6.4.1.);
- the transport of this reference mesh onto the real domain. Let $\hat{M}$ be a point in the canonical mesh with coordinates $\hat{x}$, $\hat{y}$ and $\hat{z}$ (see chapter 6); the auxiliary point $\tilde{M}$ is constructed as follows:

$$\hat{M} \Longrightarrow \tilde{M} = F(\hat{x}, \hat{y}, \hat{z})$$

where F is a function of the form already seen in chapter 6, briefly recalled below:

(a) in the case of a block with a tetrahedral topology, the transformation:

$$F(\hat{x}, \hat{y}, \hat{z}) = \sum_{i=1}^{4} P_i(\hat{x}, \hat{y}, \hat{z}) S_i$$

where:

$$P_1(\hat{x}, \hat{y}, \hat{z}) = 1 - \hat{x} - \hat{y} - \hat{z}, \ P_2(\hat{x}, \hat{y}, \hat{z}) = \hat{x},$$
$$P_3(\hat{x}, \hat{y}, \hat{z}) = \hat{y}, \ P_4(\hat{x}, \hat{y}, \hat{z}) = \hat{z},$$

and where S_i denotes the vertex i of the block,

9.3. MULTIBLOCK METHODS IN 3 DIMENSIONS

(b) in the case of a block with a pentahedral topology:

$$F(\hat{x}, \hat{y}, \hat{z}) = \sum_{i=1}^{6} P_i(\hat{x}, \hat{y}, \hat{z}) S_i$$

with:

$$P_1(\hat{x}, \hat{y}, \hat{z}) = (1 - \hat{x} - \hat{y})(1 - \hat{z}), \quad P_2(\hat{x}, \hat{y}, \hat{z}) = \hat{x}(1 - \hat{z}),$$
$$P_3(\hat{x}, \hat{y}, \hat{z}) = \hat{y}(1 - \hat{z}), \quad P_4(\hat{x}, \hat{y}, \hat{z}) = (1 - \hat{x} - \hat{y})\hat{z},$$
$$P_5(\hat{x}, \hat{y}, \hat{z}) = \hat{x}\hat{z}, \quad P_6(\hat{x}, \hat{y}, \hat{z}) = \hat{y}\hat{z}$$

(c) in the case of a block with a hexahedral topology:

$$F(\hat{x}, \hat{y}, \hat{z}) = \sum_{i=1}^{8} P_i(\hat{x}, \hat{y}, \hat{z}) S_i$$

with:

$$P_1(\hat{x}, \hat{y}, \hat{z}) = (1 - \hat{x})(1 - \hat{y})(1 - \hat{z}),$$
$$P_2(\hat{x}, \hat{y}, \hat{z}) = \hat{x}(1 - \hat{y})(1 - \hat{z}),$$
$$P_3(\hat{x}, \hat{y}, \hat{z}) = \hat{x}\hat{y}(1 - \hat{z}), \quad P_4(\hat{x}, \hat{y}, \hat{z}) = (1 - \hat{x})\hat{y}(1 - \hat{z}),$$
$$P_5(\hat{x}, \hat{y}, \hat{z}) = (1 - \hat{x})(1 - \hat{y})\hat{z}, \quad P_6(\hat{x}, \hat{y}, \hat{z}) = \hat{x}(1 - \hat{y})\hat{z},$$
$$P_7(\hat{x}, \hat{y}, \hat{z}) = \hat{x}\hat{y}\hat{z}, \quad P_8(\hat{x}, \hat{y}, \hat{z}) = (1 - \hat{x})\hat{y}\hat{z}$$

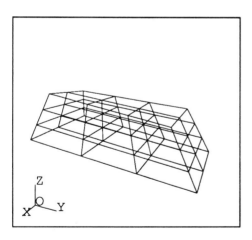
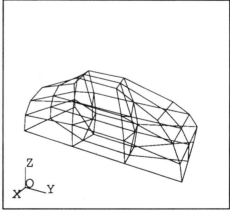

Figure 9.3: *Transport of the canonical partitioning onto the real domain.*

Using this auxiliary point $\tilde{M}$, an element deformation method is applied to create the corresponding point M (see figure 9.4 in which we show the mesh before deformation on the left-hand

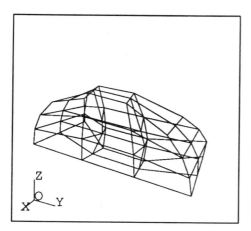
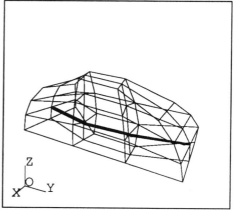

Figure 9.4: *Deformation of the elements.*

side, and the resulting mesh on the right-hand side. To clarify this figure, only one block has been considered: note that this block is one of those in figure 9.8):

$$\tilde{M} \implies M = D(\tilde{M})$$

Function D takes all the points on the faces of the real blocks into account and relocates the point which is regarded as the weighted barycentre of these points. This technique enables us to treat non-convex blocks better (function F does not guarantee that point $\tilde{M}$ is internal to the block since point M is inside the unit reference block).

$$M = \tilde{M} + \frac{1}{\alpha} \sum_{P_k \in \mathcal{F}} w_k \alpha_k D(P_k)$$

using the notation in chapter 6.

Step 3 : Enumeration of the elements resulting from the splitting. To obtain the elements constituting the mesh of the block, it suffices to enumerate their vertices (as in 9.2.1). The vertices of the elements are known via three indices, i, j and k, and the associated triplet:

$$i \quad j \quad k \Rightarrow (i, j, k)$$

9.3. MULTIBLOCK METHODS IN 3 DIMENSIONS

A global number is associated with each vertex (therefore with each triplet):

$$(i,j,k) \Rightarrow N = P(i,j,k)$$

This pointer P is constructed as seen in section 9.2.1., i.e.:

- if vertex S is a vertex of the coarse mesh, then $P(i,j,k)$ is its number in this coarse mesh;
- if vertex S is located on a coarse edge, $P(i,j,k)$ is computed according to the number of this edge in the coarse mesh (see section 13.9);
- if vertex S is located on a coarse face, $P(i,j,k)$ is computed according to the number of this face in the coarse mesh (see section 13.9);
- if S is an internal point, $P(i,j,k)$ is sequentially defined using the first free number (i.e. after the last number of the last point of the last coarse face).

P being known, it is easy to find the vertices of any element from the triplets (i,j,k) associated with the i,j,k numbering defined in the initial coarse element. To make this point clear, let us consider the case of a tetrahedral block (figure 9.5, right-hand side), as the other cases (pentahedral or hexahedral blocks) do not present any particular difficulties. The method consists of enumerating all possible situations. To do this, one layer of the splitting is considered (for example the one located between sections with indices $k = 0$ and $k = 1$). It is obvious that only four cases exist (cf. figure 9.5, left-hand side), each one depending on the position of the triangle resulting from the splitting of the face under consideration. Therefore:

(a) A triangle in position 1 serves as support for the creation of three tetrahedra (figure 9.6, left-hand side):

　i. the tetrahedron with formal vertices: i,j,k; $i+1,j,k$; $i,j+1,k$ and $i,j,k+1$.

　ii. the tetrahedron with formal vertices: $i,j,k+1$; $i,j+1,k+1$; $i+1,j,k+1$ and $i+1,j,k$.

　iii. the tetrahedron with formal vertices: $i+1,j,k$; $i,j+1,k+1$; $i,j+1,k$ and $i,j,k+1$.

132   CHAPTER 9. MULTIBLOCK METHODS

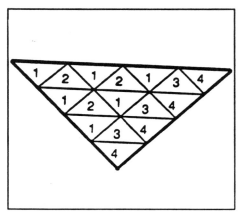 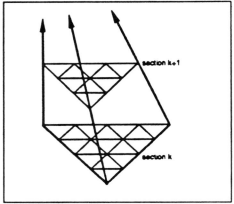

Figure 9.5: *The initial block and two sections of the splitting.*

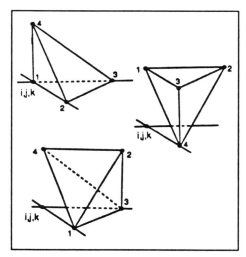 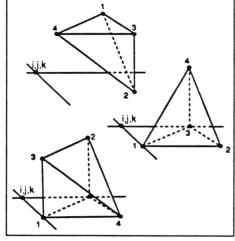

Figure 9.6: *Formal numbering (cases 1 and 2).*

9.3. MULTIBLOCK METHODS IN 3 DIMENSIONS

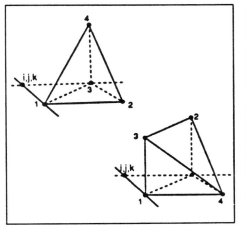
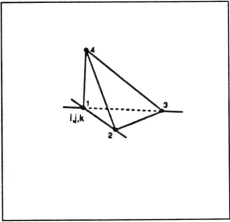

Figure 9.7: *Formal numbering (cases 3 and 4)*.

- (b) A triangle in position 2 serves as support for the creation of three tetrahedra (figure 9.6, right-hand side): :
    - i. the tetrahedron with formal vertices: $i, j+1, k+1$; $i+1, j+1, k$; $i+1, j+1, k+1$ and $i+1, j, k+1$.
    - ii. the tetrahedron with formal vertices: $i+1, j, k$; $i+1, j+1, k$; $i, j+1, k$ and $i, j+1, k+1$.
    - iii. the tetrahedron with formal vertices: $i+1, j, k$; $i, j+1, k+1$; $i+1, j, k+1$ and $i+1, j+1, k$.
- (c) A triangle in position 3 serves as support for the creation of only two tetrahedra (figure 9.7, left-hand side) (since it is placed in contact with the "skew" face of the initial tetrahedron):
    - i. the tetrahedron with formal vertices: $i+1, j, k$; $i, j+1, k+1$; $i+1, j, k+1$ and $i+1, j+1, k$.
    - ii. the tetrahedron with formal vertices: $i+1, j, k$; $i, j+1, k$; $i, j+1, k+1$ and $i+1, j+1, k$.
- (d) Finally, a triangle in position 4 is used to create one single tetrahedron (figure 9.7, right-hand side):
    - i. this is the tetrahedron with formal vertices: i, j, k; $i+1, j, k$; $i, j+1, k$ and $i, j, k+1$.

Step 4 : The pasting together of the meshes of the different blocks. This operation is implicit since the points on the faces common to two blocks are numbered globally.

Remark 9.1 : If the initial block is a segment, the resulting mesh is that of a line in space R^3. In the case of a triangular or a quadrilateral initial block, the final mesh is that of a surface in R^3. □

9.3.2 Data description

As data, this mesh generator requires a coarse discretization of the domain into linear, triangular, quadrilateral, tetrahedral, pentahedral or hexahedral type blocks to be input, as well as a distribution of points located on the edges of these blocks. The number of these points must be consistent as in dimension 2. These points will be used for splitting the coarse edges, the coarse faces and, finally, the blocks themselves.

The data of this mesh generator is schematized as follows:

$$\text{Blocks} + \text{Control points} \rightarrow \text{Mesh generator}$$

The face, edge and point references of the mesh created are derived from the face, edge and point references of the coarse mesh, whereas references of items not treated in this way are set to 0. The mesh generator associates a sub-domain number, specified by the user, with all the elements of the mesh of a given block.

9.3.3 Application example

Module COLIB2 (option NDIM = 3) of the Modulef library corresponds to the method described above. The following example results from this module. The resulting mesh (figure 9.8, right-hand side) consists of 435 elements and 658 nodes, whereas the coarse mesh (same figure, left-hand side), i.e., the set of the initial blocks, is composed of only 7 hexahedral type elements constructed from merely 28 points.

Starting with this mesh, constructed by splitting the initial blocks, we deduce the mesh of the entire domain (figure 9.9) possessing 10440 elements and 13140 vertices. To obtain this result, a series of geometrical transformations (symmetries and rotations) is processed, coupled with a succession of pasting operations. In this elaborate process for the creation of a complex mesh, we recognize the application of the methodology proposed in chapters 3 and 4.

9.3. MULTIBLOCK METHODS IN 3 DIMENSIONS

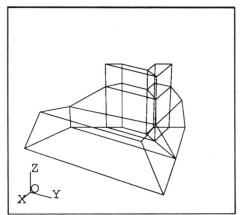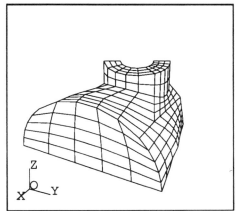

Figure 9.8: *Coarse mesh and final mesh (COLIB2).*

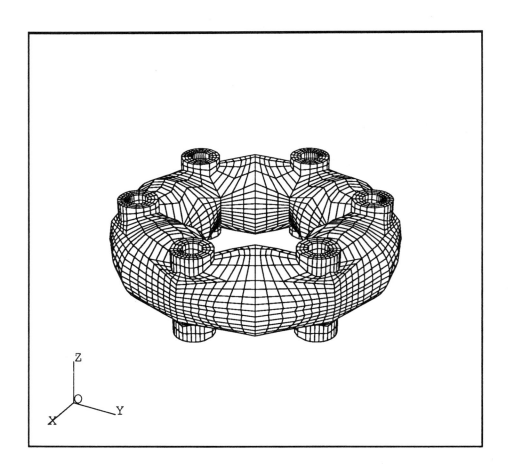

Figure 9.9: *Mesh of the entire domain after recomposition.*

Chapter 10

Advancing front methods

10.1 Introduction

This class of mesh generators, adapted to arbitrary geometries, has been studied by [A. George-1971], [Carnet-1978], [Löhner,Parikh-1988a], [Löhner,Parikh-1988b], [Peraire et al. 1988] and [Golgolab-1989]. This kind of mesh generator constructs the mesh of the domain from its boundary. The elements created are triangles in 2 dimensions and tetrahedra in 3 dimensions. The data required is the boundary, or more precisely, a polygonal discretization of it (dimension 2), input as a list of segments, or a polyhedral discretization (dimension 3), input as a list of triangular faces.

The process is iterative: a *front*, initialized by the set of items of the given boundary, is analyzed to determine a *departure zone*, from which one or several internal elements are created; the front is then updated and the element creation process is pursued as long as the front is not empty. The process can be summarized as follows (see also the scheme shown in figure 10.1):

- Initialization of the front;
- Analysis of the front:
    - Determination of the departure zone;
    - Analysis of this region:
        * Creation of internal point(s) and internal element(s);
        * Update of the front.
- As long as the front is not empty, go to "Analysis of the front".

The analysis of the front and the way in which the elements are created can be done in several ways. We propose one of these methods for the two and three-dimensional cases. The last section of this chapter introduces some extensions serving as a control of the internal point process and element creation, in such a way that the resulting mesh enjoys some particular characteristics, e.g., isotropic elements, anisotropic elements, etc.).

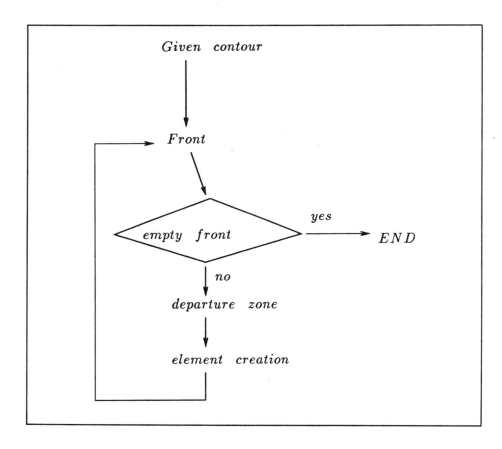

Figure 10.1: *General scheme of an advancing front method.*

10.2 Advancing front method in 2 dimensions

Before presenting an advancing front method, we should note that this type of approach represents, at least in 2 dimensions, the first automatic solution for the generation of meshes for domains with arbitrary shape.

10.2.1 Mesh generation algorithm

Suitable for arbitrary geometries, this type of mesh generator constructs the covering-up of domain Ω by triangles starting from its contour. In practice a polygonal approximation of the contour is used in terms of a list of its constitutive segments. The interior of the domain, the zone to be meshed, is well defined because of the orientation of the contour serving as data.

The initial front is defined as the set of segments of boundary C describing domain Ω.

Given front F associated with boundary C, we now detail the manner in which the triangles are created. While the process for the creation of internal triangles progresses, boundary C and front F are updated (see below). Let F be the current state of the front, then the front analysis is based on the examination of the geometrical properties of the segments that constitute it. Let α be the angle formed by two consecutive segments of front F, then three situations or *patterns* are identified (figures 10.2 and 10.3):

- $\alpha < \frac{\Pi}{2}$ (pattern a), the two segments with angle α are retained and become the two edges of the single triangle created (figure 10.2);

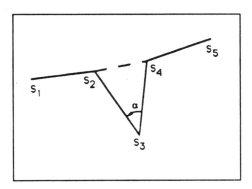

Figure 10.2: *Pattern a) and associated construction.*

- $\frac{\Pi}{2} \leq \alpha \leq \frac{2\Pi}{3}$ (pattern b), from the two segments with angle α, an internal point and two triangles are generated (figure 10.3);

- $\frac{2\Pi}{3} < \alpha$ (pattern c), one segment is retained, a triangle is created with this segment as an edge and an internal point (figure 10.3).

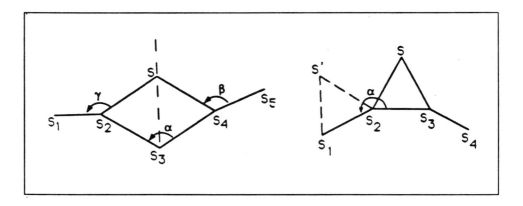

Figure 10.3: *Patterns b) and c) and associated constructions.*

The position of the internal points created is defined in such a way that they are "optimal", meaning that the elements containing these points are as regular as possible. In the case of pattern b) above, the vertex is generated on the line bisecting angle α at a distance computed from the respective lengths of the edges of the departure zone: the location of this internal point S is evaluated as follows:

$$d_{SS_3} = \frac{1}{6}(2d_{S_2 S_3} + 2d_{S_3 S_4} + d_{S_1 S_2} + d_{S_4 S_5})$$

in the case where angles β and γ (figure 10.3) lie between $\frac{\Pi}{5}$ and $2\Pi - \frac{\Pi}{5}$ radians ($\frac{\Pi}{5}$ is an empirically chosen value). For the contrary situation, the construction number 1 is used.

In the case of pattern c), a triangle, as equilateral as possible, is formed using the shortest edge of the departure zone.

At each point creation, it is necessary to verify that the point is inside the domain in its current state. This implies that any point created is, on one hand, inside the domain under consideration and, on the other hand, not located inside an existing element. This verification, crucial for this type of method, relies on exact knowledge of an adequate neighbourhood of the zone being treated (see below). In the two-dimensional case, a point will be an internal point if the intersection of all the edges resulting from it with any edge of the front is empty[1].

---

[1] In the case of domains with one or several holes, it is necessary to add the following condition: no triangles formed with the point being treated contain a point in any edge of the contour of any holes present.

10.2. ADVANCING FRONT METHOD IN 2 DIMENSIONS

Figure 10.4: *Various states of the front.*

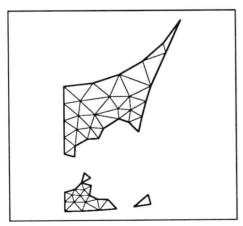

Figure 10.5: *Various states of the front (continued).*

Remark 10.1 : This analysis (which considers only three patterns) is a simplified version of a study of more complex cases as proposed in [A. George-1971]. □

A new front F is formed by suppressing:

- the edges of the present front F, belonging to a triangle created, from the present front

and by adding:

- the edges of the triangle(s) created, for edges not common to two elements, to the front.

This updated state of front F is then processed by the same method. Figures 10.4 and 10.5 show various states of the front in evolution corresponding to the domain shown in figure 10.8 (see also figures 10.6 and 10.7). Once front F is empty, the final mesh is obtained.

The efficiency and reliability of the method depend on the way the space is controlled. In practice, it is required to access the relative context of any segment in the front quickly, i.e., to determine the neighbourhood of any triangle of the mesh in evolution. This requirement, concerning the efficiency of the method, is specially important in the three-dimensional case (see section 10.3).

In the case of strongly non-convex domains, this approach may not converge. Moreover, a too sharp variation in the distribution of boundary points can produce a similar negative result. To overcome this problem, we only consider adequate primal sub-sets, or a different method must be used. In fact, this negative result is a consequence of the difficulty of proving the validity of the method theoretically, but a more astute implementation can nevertheless overcome this weakness.

Remark 10.2 : Contrary to the transport-mapping method, the number of elements of the final mesh is not calculable in advance. It is a function of the geometry, the number of boundary points and their distribution. □

Remark 10.3 : It is assumed that the boundary includes only one connected component, consequently the presence of holes requires the creation of artificial lines so that the different connected components of the boundary can be joined (see 13.4; notice, as above, that a judicious implementation can probably overcome this constraint). □

Remark 10.4 : Following the above remark, the specification of an internal line contained in the mesh is done as follows: the line is joined

10.2. ADVANCING FRONT METHOD IN 2 DIMENSIONS

to the contour and a new contour is defined by incorporating this line twice, first in one direction (in accordance with the global definition of the contour) and then in the opposite direction. □

Remark 10.5 : The triangulation obtained is clearly related to the number and relative location of the points discretizing the boundary. By specifying them, it is possible to obtain a variable density of elements in certain regions of the mesh. Some advancing front mesh generators allow the pre-specification of some desired properties (variable density of elements, anisotropic features, ..., see also remark 10.7). □

Remark 10.6 : The resulting mesh can be smoothed to obtain better quality triangles. This process corrects the position of points created using local information globally. □

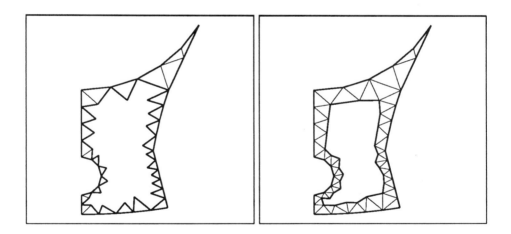

Figure 10.6: *Fronts by inflation.*

Numerous variations exist to obtain advancing front algorithms which are appreciably different. In particular, the zone of departure can be chosen as:

- a part of the contour such that its constitutive items satisfy certain conditions (the method described above falls in this group);

- the entire front, i.e., its constitutive items considered in some defined order.

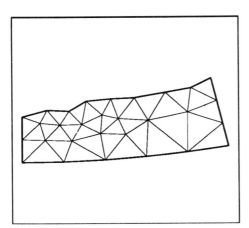 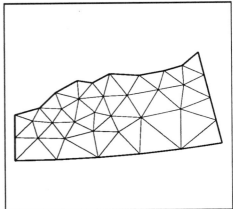

Figure 10.7: *Fronts by linear propagation.*

The first approach caters for the treatment, first of all, of particular zones, for example those with "small" angles. The second approach produces an inflation of the initial front (figure 10.6) or a propagation from an initial line (figure 10.7), the domain being, once again, that of figure 10.8 to which the reader is referred.

Remark 10.7 : The preceding discussion implicitly falls in the isotropic class where the triangles created are as close as possible to an equilateral triangle. Refer to section 10.4 for a variant of the algorithm producing anisotropic elements or elements adapted to a given criterion. □

Remark 10.8 : The advancing front technique can also be applied to create quadrilaterals. Based on the same principle, the algorithm attempts to create quadrilaterals with as regular a shape as possible. Employing empirical reasoning, this process does not prevent the creation of triangles in zones otherwise impossible to cover. □

Remark 10.9 : In order to create the internal points, [Lo-1985] proposes a different approach in dimension 2. A regular network of points is given to cover up domain Ω after which these points are connected to define the triangles. Near the boundary, a point is chosen if its location implies that it is an "optimal" point with respect to the points already present in this boundary; for the points clearly inside the domain, the connection defining the triangles is obvious. □

10.2. ADVANCING FRONT METHOD IN 2 DIMENSIONS

10.2.2 Data description

Concerning the data, this algorithm requires the input of a polygonal discretization of the contour of the domain under consideration. Following the Modulef approach, the contour is formed by the union of characteristic $\mathcal{L}$ines, in turn defined by characteristic $\mathcal{P}$oints. The data of the first contour line and the direction in which it runs, determine without ambiguity both the interior of the domain and the zone to be covered up.

The data of these kinds of mesh generators is summarized as follows:

$$\mathcal{P} \rightarrow \mathcal{L} \rightarrow \text{Contour} \rightarrow \text{Mesh generator}$$

The references of the edges and the contour points of the mesh created are derived from those of the characteristic $\mathcal{L}$ines and their items (segments and intermediary points) and characteristic $\mathcal{P}$oints; references not dealt with in this way are set to the value 0. A sub-domain number, specified by the user, is assigned to all elements by the mesh generator.

10.2.3 Application examples

Taking the example in section 6.4.3, we obtain the mesh shown in figure 10.8. From the 9 characteristic $\mathcal{P}$oints (figure 6.12), the 9 characteristic $\mathcal{L}$ines are created which describe the contour of the domain. Mesh generator TRIGEO of the Modulef library, based on the method discussed in this section, generates the mesh of the domain in triangles. To illustrate the sensitivity of the method with respect to the data, two sets of data are input containing a different number of points and different point distributions. The resulting meshes are shown in figure 10.8.

The mesh (figure 10.8) on the left-hand side contains 148 triangles and 92 nodes, that on the right-hand side contains 186 triangles and 118 nodes. For the latter, a smoothing process seems to be appropriate (see section 13.6 and figure 10.9).

In addition, it should be mentioned that this algorithm (at least as implemented in the Modulef library) does not converge in the example in figure 11.8 (left-hand side); such a primal sub-set is too complex for this method. Two solutions exist to overcome this difficulty: the modification of the number and location of the contour points or an artificial splitting of the domain into primal sub-sets of simpler shapes. Figure 10.10 shows the state of the mesh at the moment the algorithm is no longer able to progress. The problem seems to be the encounter between two incompatible

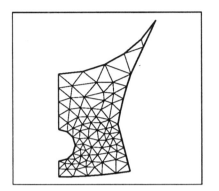 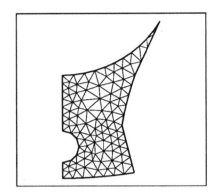

Figure 10.8: *Applications (TRIGEO) with two data sets.*

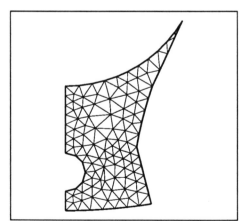 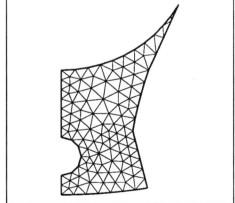

Figure 10.9: *Mesh before and after smoothing.*

10.3. ADVANCING FRONT METHOD IN 3 DIMENSIONS

fronts. [Golgolab-1989] proposes the removal of the awkward elements in this encounter zone and the restart of the method while slightly modifying certain internal parameters as a possible solution. In fact, the problem can be analyzed as the result of the creation of badly located points which obstructs the process in progress at an ulterior iteration.

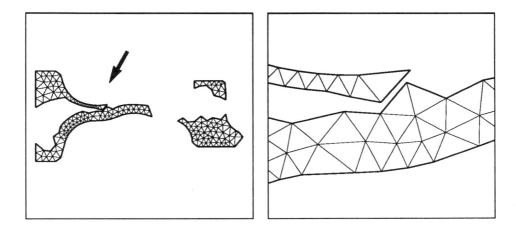

Figure 10.10: *Algorithm failing and zone of collapse.*

10.3 Advancing front method in 3 dimensions

The application of advancing front techniques in the three-dimensional case is obviously more delicate, the problems inherent in this type of approach are more difficult to solve. In this respect, the reader is referred to [Peraire et al. 1988], [Löhner,Parikh-1988b] and, more recently, to [Golgolab-1989] who proposes some efficient solutions.

10.3.1 Mesh generation algorithm

Suitable for arbitrary geometries, a mesh generator of the present class constructs the covering-up of domain Ω by tetrahedra from its boundary data. In practice, a polyhedral approximation of the boundary is used which consists of a list of its constitutive triangular faces. The interior of the domain, namely the zone to be meshed, is determined through the orientation of the data.

The algorithm is based on the same type of scheme as in dimension 2 (figure 10.1). The process is iterative: from a given contour C and associated front F, the properties of the faces of F (in terms of size and angle), as well as their neighbourhoods, are analyzed in order to select a departure zone.

According to the case being considered, either an element is created with the selected faces, or an internal point[2] is generated which allows the creation of the elements by simply joining it to the selected faces.

A new state of front F is formed by removing:

- the faces of front F, belonging to a tetrahedron created, from the present front.

and by adding:

- the faces of the tetrahedron(a) created, in the case where these faces are not common to two elements, to the present front.

This updated front F is then processed in the same way. Once it is empty, the final mesh is obtained.

The efficiency and reliability of the method are strongly related, even more than in the two-dimensional case, to the way in which the space is known. It corresponds to finding the context relative to a face of the front quickly, i.e., to known the neighbourhood of any triangle of this front and that of any tetrahedron of the mesh in progress.

10.3.2 Data description

The data required for this mesh generator consists of a discretization of the contour of the domain under consideration. This contour is formed from a mesh composed of triangular faces. This mesh can, for example, be produced with the help of a CAD system (cf. chapter 12).

The data relevant to this mesh generator is the following:

Mesh of the contour → Mesh generator

---

[2] See the two-dimensional case; here in three dimensions, the condition regarding the intersection of the edges is replaced by the two following conditions: on one hand, the edges resulting from this point do not intersect a face of any element of the front and, on the other hand, the faces containing this point are not intersected by an edge of the front; the other condition in the case of a domain with hole(s) remains valid.

10.3. ADVANCING FRONT METHOD IN 3 DIMENSIONS

The face, edge and boundary point references of the mesh created are derived from those of the *surface* mesh input as data. The references of the items not affected in this way are set to 0. The mesh generator associates a sub-domain number, provided by the user, with all elements.

10.3.3 Application examples

The mesh generator DARIUS, developed at INRIA [Golgolab-1989], corresponds to this method and is used to produce the following examples.

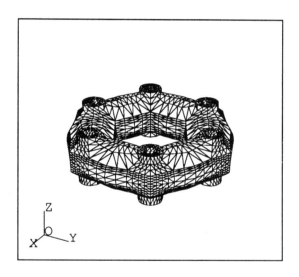

Figure 10.11: *Mesh of a junction (DARIUS)*.

Figure 10.11 displays an initial example of an application of this generator. It corresponds to the mesh of a part of a junction (this domain is also shown in figure 9.9, to which the reader is referred). The data required by the advancing front generator is the surface mesh of the boundary of the domain in the form of a list of its triangular faces; this data contains 5760 triangles and 3486 points.

The mesh of the interior of the object includes 17107 tetrahedra. Figure 10.11 shows the visible faces of this mesh.

The advancing front approach is also suitable for the mesh generation of the exterior of objects (both in $2D$ and $3D$). In this case, the data consists of a surface mesh of the object and that of a surface bounding the exterior of the zone of computation. The example displayed in figure 10.12 shows

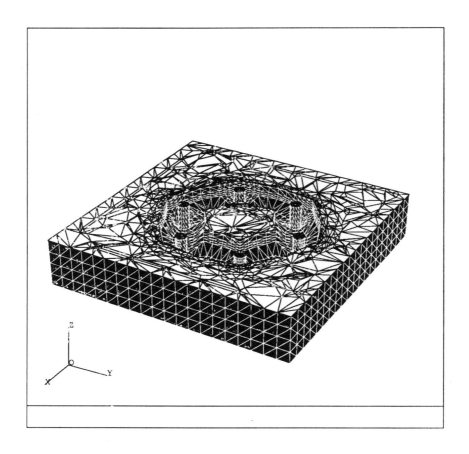

Figure 10.12: *Mesh of the exterior of the above junction (DARIUS).*

10.3. ADVANCING FRONT METHOD IN 3 DIMENSIONS

a domain bounded by the junction (figure 10.11) and a parallelepipedal "box" enclosing it; the cut enables us to examine a part of the result; the latter includes 30187 elements.

Remark 10.10 : For an advancing front method by inflation (see above), the enclosing "box" is not necessarily defined. In this case, the generator can be stopped when the elements created are at a given distance from the object under consideration. The set of external element faces of the last layer constitutes the external boundary of the meshed zone. □

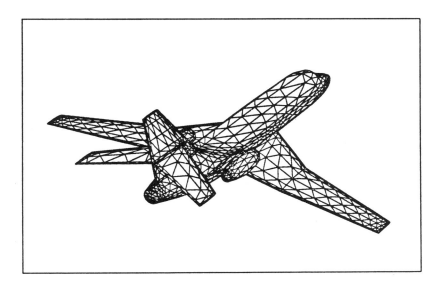

Figure 10.13: *Surface mesh (data)*.

The second example is a cut of a mesh of a Falcon type aircraft exterior. The domain of interest is, in this case, the region bounded by this object and a surface where the farfield boundary conditions can be applied. This surface has been chosen (for a comprehensive illustration) as a sphere which is far enough from the body of the aircraft. The data (provided by DA.[3]) includes 4722 triangles and 2361 points; the final mesh created by module DARIUS, consists of 39015 tetrahedra. Figure 10.13 displays the data, i.e., the surface mesh of the aircraft, while figures 10.14 and 10.15 show, respectively, a cut of the mesh created around this aircraft and a zoom into this cut.

---

[3]DA. : Dassault - Aviation

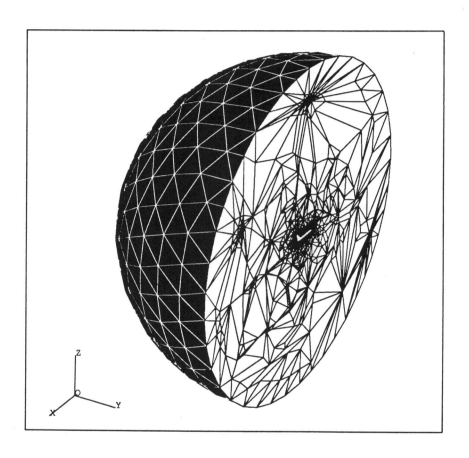

Figure 10.14: *Mesh around the above aircraft (cut through the result).*

10.3. ADVANCING FRONT METHOD IN 3 DIMENSIONS

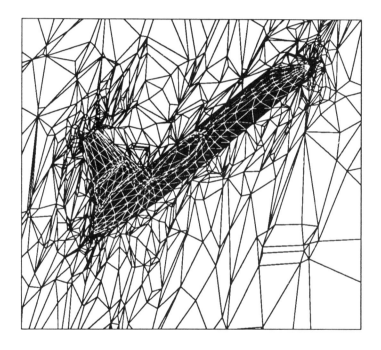

Figure 10.15: *Mesh around the aircraft (zoom into the above cut).*

10.4 Extensions

The following points are discussed briefly in this section:

- The definition of a neighbourhood space which provides an easy and fast access to elements neighbouring a given element,

- The definition of a control space with the purpose of governing the internal point creation phase,

- The effective use of this space for the creation of internal points and associated elements.

All these ideas and the implementation of the corresponding algorithms on computer must take into account the fact that the transfer of an operation whose nature is purely geometric to computer is only a more or less exact approximation of this operation (see also 11.6.4).

10.4.1 Neighbourhood space

One of the solutions for finding the neighbourhoods of a given element quickly is to create a grid enclosing domain Ω, the boxes of this grid referencing the useful information. This space, called a *neighbourhood space*, corresponds to the following definition:

Definition 10.1 :
$\mathcal{V}$ is called a **neighbourhood space** for mesh T of domain Ω if:

- $\Omega \subseteq \mathcal{V}$

- the elements of $\mathcal{V}$ "include" the neighbourhood of any element K of mesh T. □

From a geometrical point of view, $\mathcal{V}$ is a partitioning enclosing Ω; the easiest step seems to be the choice of a regular grid, or a grid of type "quadtree" in dimension 2 or "octree" in dimension 3 (see section 8.3). This structure is created in such a way that its boxes reference the edges (faces) of elements K of T intersected by it (in practice, a box refers to a table or a list).

The neighbourhood space is initialized as follows: associate the information relative to the neighbouring items with each of its boxes containing one or several items (point, edge and face). The contents of the space are then updated at each element creation.

10.4. EXTENSIONS

Via this structure, the set of the neighbouring elements can immediately be known for any element of the front to be considered, as well as the set of neighbouring edges (faces) and neighbouring points. Consequently, the front analysis is easy and rapid and the environment of the zone being processed is known.

10.4.2 Control space

To govern the internal point creation and connected element creation phase, a *control space* is introduced. This is an easy way to know the present environment of any zone in the space. In fact, when creating an element from the current state of the front, other information is not known a priori other than that relative to the portion of the front close to the departure zone being processed. In particular, it is not known what happens ahead of the front, a fortiori far from the zone we are in (roughly speaking, we are in the "fog").

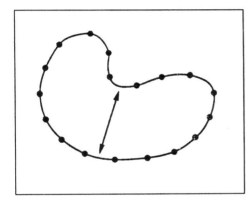
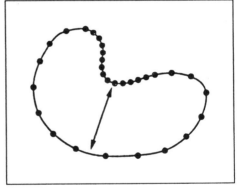

Figure 10.16: *Two types of discretization of the same boundary.*

In figure 10.16, two types of data are shown to indicate the nature of the constraints present in the given direction for the $2D$ case. These constraints are linked to the discretization of the portion of the front opposite to the departure zone under consideration. On the left-hand side of the figure, data is displayed which induces the creation of triangles with a regular shape and size naturally. The data on the right-hand side indicates that a finer mesh must be generated in the zone close to the portion of the contour which is discretized more finely, while a coarser mesh is required elsewhere. It corresponds therefore to an adequate management of these two imperatives in such a way that the desired progressivity is obtained.

In order to solve this problem, the above mentioned control space is introduced as follows:

Definition 10.2 :
(Δ, H) is a **control space** for mesh T of domain Ω if:

- $\Omega \subseteq \Delta$ where Δ is a covering-up enclosing Ω

- With each $P \in \Delta$ a functional $H(P, d)$ is associated, where d is a direction of sphere S^2 (of circle S^1 in dimension 2):

$$H(P, d) : \Delta \times S^2 \to R \qquad \square$$

The functional H, lying on covering-up Δ, allows the specification of properties to be satisfied or that of desired criteria to be verified by the elements of the desired mesh.

From a geometrical point of view, Δ is any type of partitioning, for example, one of the following types:

a) a partitioning of the type "quadtree" in dimension 2, or "octree" in dimension 3 (cf. section 8.3),

b) a regular finite difference type partitioning,

c) any mesh created by the user.

Besides its covering-up aspect, (Δ, H) includes, via functional H, global information about the physical nature of the problem. The encoded values serve to verify if mesh T, under construction, satisfies the functional at all points P (see the next section).

To construct H, one of the following approaches can be followed:

- compute the local h's of the given points from the data (local h can be chosen as the distance desired between points) by, for example, a generalized interpolation from which H is deduced (this is a purely geometrical procedure in the sense that it relies on the geometrical properties of the data: size of the edges or faces of the contour, ...);

- define H manually for each element of partitioning Δ (for example, provide the desired stepsize in the space - *isotropic* control - or along a direction - *anisotropic* control);

10.4. EXTENSIONS

- specify H manually by specifying its value for each element of a covering-up created for this purpose (this is the case for a space of type c) introduced above);

- in case a) above, the box sizes can serve to encode the value H, resulting from the construction of an appropriate splitting.

In addition, the control space provides a global vision of the space so that the above mentioned "fog" effect disappears. Thanks to the functional, H, an encoding of the useful information is given. When considering the zones of departure in the examples shown in figure 10.16, functional H includes, in a consistent manner, some information regarding the zone to which we are progressing. Consequently, the junction between elements created using this zone of departure and elements arising from the opposite zone is managed smoothly.

Remark 10.11 : This control space notion is defined similarly in the Voronoï type methods in section 11.6.2. □

10.4.3 Control and internal points

Starting from current state of the mesh and thus the associated front, (Δ, H) is used to obtain the new state of this mesh by creating elements verifying the properties contained in the control space.

An item (face in dimension 3, edge in dimension 2) of the front is considered, via the neighbourhood space: its environment is known and, via the control space, the nature of the elements to be created in the zone is determined. In particular, the value of functional H provides the information relative to the desired shape of the elements under construction directly and, in fact, H provides a means to specify both a direction to be followed and a distance to be respected.

Several cases are encountered:

- Generation of an element without the creation of internal points. A point verifying H already exists in the neighbourhood of the item of the front, and it therefore suffices to link the face (edge) to this point,

- Creation of an internal point and generation of an element including it as a vertex. Functional H defines the optimal location of the point to be created, after which it is joined to the face (edge) being processed,

- locked situations. Among these, the following are encountered:

- the point verifying H is created outside domain Ω,
- this point is located in an existing element (collision).

In any of these cases, the algorithm must provide some flexibility in order to satisfy, at best, the present constraints. In particular, the deletion of a set of existing elements may be required so that the environment is modified and the difficulty is overcome [Golgolab-1989].

Remark 10.12 : In practice, the mesh will verify the required properties in an approximate manner, i.e. as long as an impossible case is encountered. In this case, one has to take the constraints which are acting in opposite directions into account carefully in such a way that an average solution is reached. □

To conclude, we note that, at least in the three-dimensional case, the implementation of an efficient and robust advancing front method suitable for the creation of elements satisfying the desirable criteria is a relatively complex task which requires great care. This is the reason why only a few advancing front mesh generators are really efficient in dimension 3.

Chapter 11

Voronoï methods and extensions

11.1 Introduction

Mesh generators currently called Voronoï mesh generators[1] have this denomination because they use a point insertion process which leads to Delaunay meshes (cf. 11.2) associated with the set of these points[2].

Numerous authors, among which are [Hermeline-1980], [Watson-1981] in the early 80s and then [Jameson et al. 1984], [Cavendish et al. 1985], [Cendes et al. 1985], [Cendes,Shenton-1985], [Baker-1986], [Coulomb-1987], [Perronnet-1988a], etc. have investigated mesh generators based on these types of methods. In practice, the various approaches depend strongly on the way the boundary of the domain under consideration is dealt with. In fact, two classes of methods can be enumerated:

- those incorporating a *global definition* of the boundary: they produce a mesh respecting the boundary geometry by generating, if necessary, new points on it;

- those based on a *discretization* of this boundary where the constraint is to maintain the integrity of the data (i.e. this approximation must appear exactly in the resulting mesh).

The latter approach, which is more constraining and rarely explicitly mentioned (at least in the three-dimensional case), permits a modular con-

---
[1] We retain this name for simplicity.
[2] From any Delaunay mesh, an associated Voronoï mesh can be created by joining the centrepoints of the circumballs connected to the elements sharing a common vertex.

ception of meshes. In this study, one of the ways of constructing a mesh generator satisfying this constraint will be illustrated.

After recalling the point insertion process mentioned at the beginning of this chapter, we will show that it can be employed for the purpose we are dealing with. The associated difficulties will be pointed out and some possible solutions will be given.

11.2 Outline of the Delaunay-Voronoï method

Among the numerous ways of presenting this method, we choose the one which is the most suitable in terms of the desired objective: the derivation of a general application mesh method. For simplicity, the 2 and 3 dimensional cases are presented (in fact, this method can be extended to all dimensions).

Let $x_1, x_2, ..., x_n$ be a set of n distinct points in the plane or space, denoted by $\{x_k\}$, and T_i a mesh whose element vertices are the first i points of $\{x_k\}$, then mesh T_{i+1} will be derived from mesh T_i in such a way that point x_{i+1} of the initial set is one of its vertices.

From the two following sets:

- $\mathcal{B}_i = \bigcup_j B_j$ where B_j is the circumcircle (the circumsphere) of element K_j of T_i

- $T_i = \bigcup_j K_j$ is the triangulation including the first i points

it is possible to characterize the location of point x_{i+1} exactly.

In fact, it is obvious that only three situations are possible (for simplicity figure 11.1 presents the two-dimensional case):

- a) $x_{i+1} \in T_i$ i.e., $\exists K_j \in T_i$ such that $x_{i+1} \in K_j$,

- b) $x_{i+1} \notin \mathcal{B}_i$ i.e., x_{i+1} is not in any circle (sphere) B_j,

- c) $x_{i+1} \notin T_i$ but $x_{i+1} \in \mathcal{B}_i$, i.e., x_{i+1} is not in an element K_j of T_i but is in a circle (sphere) B_j.

A constructive method of creating the desired mesh T_{i+1} is associated with each of these configurations. A theoretical analysis of this construction can be found in [Hermeline-1980]. In the remainder of this section, the two-dimensional case is investigated, and can be easily extend to the three-dimensional case while replacing notions of circles and edges by those of spheres and faces.

Let us denote a point x_{i+1} by x.

11.2. OUTLINE OF THE DELAUNAY-VORONOÏ METHOD

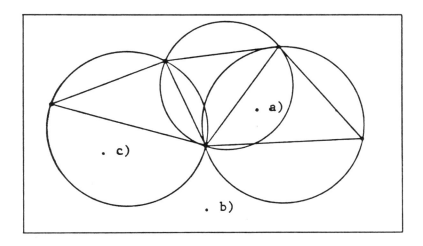

Figure 11.1: *The 3 possible configurations.*

1. Situation a): Let S be the set of elements of T_i whose circumcircle contains point x and let $F_1...F_p$ be the external edges of this set, then we can show that these faces are "visible" from point x[3]. Thus, the following holds:

   $T_{i+1} = (T_i - S) \bigcup \{F_j, x\}_j$ $1 \leq j \leq p$. Mesh T_{i+1}, obtained in this way, includes the first $(i + 1)$ points of set $\{x_k\}$ as element vertices (figure 11.2). The new elements, denoted by $\{F_j, x\}$, are created by joining point x to the external edges F_j in such a way that their surface is positive (cf. section 2.3.4.).

2. Situation b): As $x \notin B_i$, $x \notin T_i$, we look for $F_1...F_p$, the external edges of T_i, which define a half-plane strictly separating x and T_i, then:

   $T_{i+1} = T_i \bigcup \{F_j, x\}_j$ $1 \leq j \leq p$ is the Delaunay triangulation associated with the first $(i+1)$ points of set $\{x_k\}$ (figure 11.3 left-hand side). As above, the new triangles must be enumerated correctly.

3. Situation c): Let S be the set of elements of T_i whose circumcircle contains x and let $F_{i_1}...F_{i_q}$ be the edges not common to two elements of S defining a half-plane not separating x and T_i, and, among the external edges of elements of T_i not already dealt with, let $F_{i_{q+1}}...F_{i_p}$ be those defining a half-plane separating x and T_i, then:

---
[3]The set S is star-shaped with respect to x.

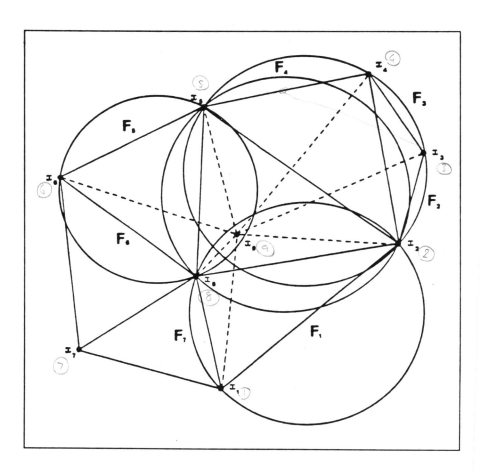

Figure 11.2: *Situation a) : insertion of point x_9 (by Hermeline)*.

11.2. OUTLINE OF THE DELAUNAY-VORONOÏ METHOD

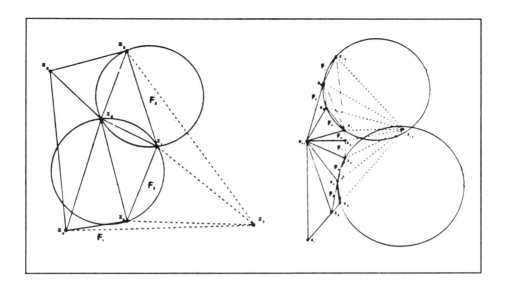

Figure 11.3: *Situations b) and c) : insertion of points x_7 and x_{12} (ibidem).*

$T_{i+1} = T_i - S \bigcup \{F_{i_j}, x\}_j$ $1 \leq j \leq p$ is the Delaunay triangulation including the first $(i+1)$ points of set $\{x_k\}$ (figure 11.3 right-hand side).

The speed of the process is a function of different parameters linked to:

- the search for element(s) containing the point to be inserted;
- the search for elements to be removed;
- the way the points of the set are introduced (the order of insertion). In fact, the latter influences the cost of these searches.

The fundamental property of this construction is that the triangulation obtained is of the type *Delaunay*. There exist several characterizations of this type of triangulation; among them, are:

- The circumcircle associated with each triangle does not include any other points of the set (cf. figure 11.4, right-hand side). This property remains true whatever the dimension of the space; in this manner, in 3 dimensions, each circumsphere associated with a tetrahedron is empty[4];

---

[4]This property is often called the "criterion of the empty sphere".

- In 2 dimensions, each pair of triangles having a common edge is such that the smallest of the angles formed by two consecutive edges is maximal[5] (cf. same figure, left-hand side: from four given points, the Delaunay configuration is that where the angles are denoted by α_i; the other mesh possible (where angles are denoted by β_i) is not of type Delaunay);

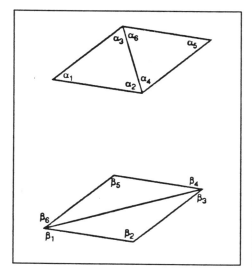
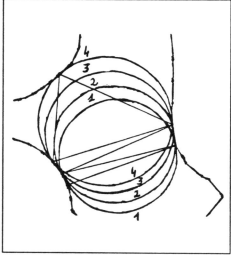

Figure 11.4: *Some properties of Delaunay triangulations.*

Consult [Cherfils,Hermeline-1990] who consider the two-dimensional case and, more generally, [Preparata,Shamos-1985] who list the properties of Delaunay type triangulations.

11.3 Mesh generation application

The application of this point insertion process makes the construction of a mesh of the convex envelope of the initial set of points possible. In the case of non-convex domains defined by their boundary, it is not possible to obtain a mesh *respecting* the corresponding domain in this way. To clarify this point, we will give some definitions related to this notion of *respect*.

---

[5]What is the extension of this property, for instance in three dimensions ?

11.3. MESH GENERATION APPLICATION

We consider the usual case where the domain to be meshed is defined by the boundary data. Two major classes of definition are currently used:

- Type 1 : *Global definition* of the boundary: it is known analytically or parametrically (equations, splines, Bézier patches, etc., cf. chapter 12);

- Type 2 : *Discrete definition* of the boundary: it is known via the list of its edges in 2 dimensions and, in 3 dimensions, via the list of triangular faces which compose its mesh.

Another possibility for defining the boundary of a domain is to combine these two approaches.

Within the framework of this study, we consider a definition of type 2 which corresponds to the data which we have at our disposal. The subjacent constraint can be formalized by the following definition:

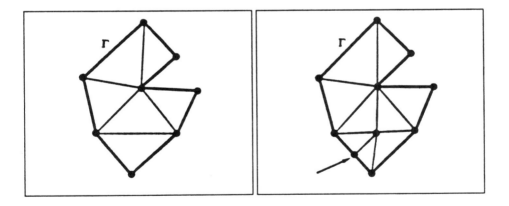

Figure 11.5: *Respect of a given boundary.*

Definition 11.1 : $\mathcal{T}$ is a mesh of Ω *respecting* boundary Γ of this domain if the following conditions hold:

- $\overline{\Omega} = \cup_{K \in \mathcal{T}} K$;

- The given boundary, i.e., all its constitutive items (point, edges and faces) exist in $\mathcal{T}$. □

Figure 11.5 shows two meshes of the same domain. If the data description of the boundary is constituted of the eight external edges of the left-hand side triangulation, the right-hand side mesh, though valid, does not respect the data. In fact, a point has been created on one of the initial edges and consequently, this edge no longer exists in the resulting mesh.

On the other hand, independently and even in the case of convex domains, the mesh produced using only the boundary data does not contain, a priori, any internal points and is consequently unsuitable for computation.

The construction of a mesh generator based on the point insertion process, as described, poses therefore the three following problems:

- The appropriate use of the point insertion process;
- The respecting of boundary Γ of the domain according to definition 11.1;
- The construction of the set of internal points.

The purpose of the sections below is to propose some possible answers to deal with these problems. Firstly, the two-dimensional case is examined after which the three-dimensional case is presented. The latter is obviously more delicate and numerous studies are presently in progress in this field.

11.4 The two-dimensional case

11.4.1 Mesh with specified vertices

For a domain defined by the list of the boundary edges, we propose the following mesh generation scheme:

1. Creation of the set of points associated with the data, i.e., the endpoints of the edges of the boundary of the domain.

2. Determination of the position of four supplementary points such that the quadrilateral formed by these includes all the points of the set.

3. Creation of the mesh of this quadrilateral using two triangles (let T_0 be this mesh).

4. Insertion of the points, one by one, in the set to obtain a mesh containing these points as element vertices.

11.4. THE TWO-DIMENSIONAL CASE

To perform this step, the above insertion process is used (cf. section 11.2) where a new mesh is created from a given mesh and an internal point in such a way that this point is an element vertex. Then denote:

- T_n a triangulation including the first n points of a set as vertices (T_0, the initial covering-up of the enclosing quadrilateral initializes T_n),
- x_{n+1} the next point in this set.

According to phase 3 of the above scheme, point x_{n+1} is inside T_n. More precisely, only two situations are possible:

(a) x_{n+1} is inside an element K_i of T_n,
(b) x_{n+1} is on the edge common to two elements, K_j and K_k, of T_n.

The third case possible would be that where x_{n+1} is identical to one of the points already present in the mesh; this case is excluded as the points in the set specified are assumed distinct.

From K_i (or from the elements K_j and K_k), set S is created by a tree search. This is the set of elements of T_n such that:

- x_{n+1} is internal to the circumcircles associated with the elements of S.

The construction of triangulation T_{n+1} is then done in the same way as seen above in case a):

- include the elements of T_n not included in S in T_{n+1},
- remove the elements in S and remesh this set by joining point x_{n+1} to the external edges of S.

Refer to figure 11.2 for an illustration of the insertion of a point in the case corresponding to the hypothesis assumed (preliminary creation of mesh T_0).

Once all the points of the initial set have been introduced according to this method, the covering-up of the initial quadrilateral by triangles is obtained. This mesh is called the "box" mesh (figure 11.6 left-hand side):

Definition 11.2 : A *box* mesh of a domain Ω is a triangulation which covers a quadrilateral enclosing the domain, and whose element vertices are:

the points specified and

the 4 points defining the enclosing quadrilateral. □

In practice, the enclosing quadrilateral is chosen sufficiently far from the object under consideration.

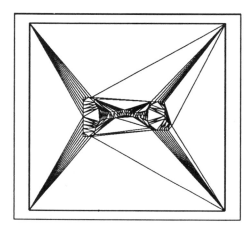 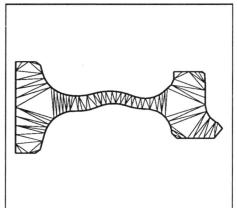

Figure 11.6: *"Box" mesh and "boundary" mesh.*

From the box mesh, a mesh called the boundary mesh is constructed (figure 11.6 right-hand side), defined as follows:

Definition 11.3 : The *boundary* mesh of a domain Ω is a triangulation constructed solely from the boundary points of Ω, in such a way that definition 11.1 holds. □

A priori, this mesh does not include any point internal to Ω according to 11.4.3 (see this section). In particular, in 2 dimensions, the vertices of this mesh consist only of the boundary points of the domain. In 3 dimensions, we will see that it can include a few internal points.

Supposing that the box mesh contains the given boundary exactly, then one can suppress all the exterior triangles by taking the location of the elements with respect to this boundary into account. Consequently, the desired boundary mesh is obtained.

Remark 11.1 : By eliminating all the elements with a vertex which is one of the 4 points created to define the enclosing box from the box

11.4. THE TWO-DIMENSIONAL CASE

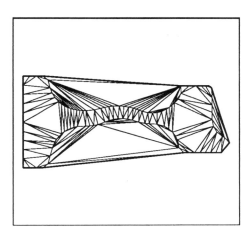

Figure 11.7: *Mesh of the envelope of a set of points.*

mesh, the mesh of the convex envelope of domain Ω is obtained (figure 11.7 corresponds to a case where the envelope obtained is not convex cf. [George,Hermeline-1989]). □

11.4.2 Respecting a prescribed boundary

To create the boundary mesh it is necessary to define the notions of interior and exterior with respect to the domain considered.

The notion of interior and exterior is only defined with respect to the boundary of the domain and therefore with respect to the segments of the polygon approximating this boundary, where this polygon is the data supplied by the user. Consequently, it is only meaningful if the box mesh includes all these segments in the list of its edges.

Regularity conditions for data corresponding to this case are discussed in [Hermeline-1980] and [George,Hermeline-1989]. A method for the creation of a mesh satisfying this crucial property, even in the case where the data is not "admissible", is described in [George et al. 1989a].

It seems that, at least in two dimensions, a data set not ensuring this property is not suitable for the creation of a good quality mesh with respect to the finite element method. We suggest that the user provides a more accurate approximation of the boundary in this case (to do this, it suffices to add points on the critical edges).

In practice, a mesh generator based on the method discussed in this section must take this problem into account carefully and recover from bad situations (see the three-dimensional case) or, at worst detect them and provide a directly usable diagnostic.

11.4.3 Construction of the final mesh

To obtain the final mesh, it is necessary to proceed to the creation of the internal points. Several approaches are suitable, among which we only mention two:

- Insert a point in each element which is too large with respect to a value computed from the local h associated with the vertices (the local stepsize, h, is a measure, for example, of the desired length of the edges having the vertex under consideration as an endpoint) [Hermeline-1980]. Another solution is to insert a point inside each element whose circumcircle is "large" [Holmes,Snyder-1988]. The location of the internal point to be created is computed by a weighted barycentre (cf. section 13.6).

  Some solutions are found, particularly, in the above mentioned references, which allow us:

  - to decide if a point must be created inside an element,
  - to find its location,
  - to assign a local h to it by a geometric, arithmetic, harmonic average, etc.

- Evaluate each element in the mesh by computing its quality (i.e. its degeneracy), its surface and the consistency between the length of its edges and the local h's of its vertices. If the element fails in one or other of these criteria, the optimal point, associated with its smallest edge, is created. If this point is inside the element or is located inside one of its neighbours, the point is inserted; if not the next triangle is considered.

Other methods exist to deal with this problem, for example, in section 11.6.2, a different and more global approach is introduced to provide a control of the internal point insertion process. The so-called internal point insertion is done by using the insertion process described[6] in section 11.2 in the case of situation a).

---

[6]See also section 11.5.3.

11.4. THE TWO-DIMENSIONAL CASE

Remark 11.2 : The triangulation obtained by this method is a Delaunay mesh. The produced connectivity is optimal[7] with respect to the set of points considered. □

Remark 11.3 : The produced triangulation is clearly dependent on the number and relative position of the points discretizing the boundary. By specifying them adequately, a variable density of elements in certain regions can be obtained. □

Remark 11.4 : The resulting mesh can be smoothed (cf. chapter 13). Although optimal (see remark 11.2), the mesh can be improved by moving its internal points. □

This is the most general mesh generator we can define. Its applications are numerous. In particular, it can be used for a large range of purposes by modifying different options (obtain the box mesh or boundary mesh only, pre-specify all or some of the internal points, obtain the mesh of the envelope of a set of points (i.e. the boundary is not given, figure 11.7), etc.).

As an illustration, the time required for the mesh creation using this method is now displayed. The time was measured on an Apollo DN4000:

Number of vertices	58	540	1140	1740	2180	2400
CPU time (in sec.)	10	18	28	37	55	62

Table 11.1 : *Required time as a function of the number of vertices.*

Numerous authors have computed the theoretical complexity of methods of this type. This measure is only an average indication as it only corresponds to the point insertion process and does not consider the problem of maintaining the integrity of the edges (edges and faces in three dimensions) specified initially.

11.4.4 Data description

As data, this mesh generator requires the input of a polygonal discretization of the boundary of the domain of interest. To illustrate this point, we follow the approach used in the Modulef code: the boundary is composed of the union of characteristic $\mathcal{L}$ines in turn defined by the characteristic $\mathcal{P}$oints. The data corresponding to the first boundary line, described in an anticlockwise direction, and the direction in which it runs determines

---
[7] According to the Delaunay triangulation properties.

without ambiguity the interior of the domain, and thus the region to be meshed.

For domains containing some holes (i.e. whose boundary has several connected components), we provide, firstly, the lines of the exterior boundary in an anticlockwise direction, and then the lines of the boundaries of the holes in a retrograde direction.

In the case where the data only consists of a set of points, the mesh of the envelope, generally the convex envelope, of this set is obtained ([George,Hermeline-1989]).

Moreover, one can specify some internal points which will be vertices in the final triangulation.

To summarize, the data for this mesh generator can be schematized as follows:

$$\mathcal{P} \to \mathcal{L} \to \text{Boundary} \to \text{Mesh generator}$$

or

$$\mathcal{P} \to \text{Mesh generator}$$

in the particular case where the mesh of the envelope of the set is desired.

The edge and boundary point references of the created mesh are derived from the characteristic $\mathcal{L}$ines and their constitutive items (segments and intermediary points) references and, finally, from the characteristic $\mathcal{P}$oint references; those not dealt with in this way are set to 0. The mesh generator will associate a sub-domain number, specified by the user, with all the elements in the mesh.

11.4.5 Application examples

Module TRIHER of the Modulef library corresponds to this method and is used to illustrate one of its numerous applications.

Figure 11.8 (left-hand side) uses the example shown in section 6.4.2. and shows the final mesh obtained after the creation of the internal points. It corresponds to a cut of a wheel of a railway vehicle, the triangulation of which consists of 550 triangles and 332 vertices. The right-hand side of this figure shows another mesh example in the case of a domain containing holes.

11.5. THE THREE-DIMENSIONAL CASE

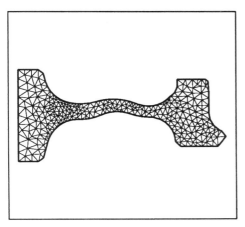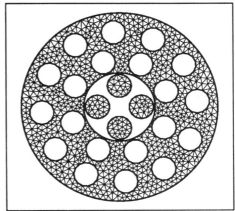

Figure 11.8: *Final meshes (TRIHER)*.

11.5 The three-dimensional case

11.5.1 Mesh with specified vertices

For a domain defined by a list of triangular boundary faces, we propose the following mesh generation steps:

1. Creation of the set of points associated with the data, i.e., the boundary points of the domain (the endpoints of the triangles discretizing this boundary).

2. Calculation of the location of eight supplementary points in such a way that the hexahedron formed by these points contains all the points in the set.

3. Creation of the mesh of this hexahedron using five tetrahedra (denote T_0 to be this mesh).

4. Insertion, one by one, of the points of the set to obtain a mesh including these points as its element vertices.

    To perform this step, we use the above insertion process (cf. section 11.2) in which a new mesh is created from a given mesh containing an internal point in such a way that this point is an element vertex. So, denote by:

- T_n a triangulation including the first n points of a set as vertices (T_0, the initial covering-up of the enclosing hexahedron, initializes T_n),
- x_{n+1} the next point in the set.

According to phase 3 of the above scheme, point x_{n+1} is inside T_n and, more precisely, only three situations are realizable:

(a) x_{n+1} is inside an element K_i of T_n ,

(b) x_{n+1} is on the face common to two elements, K_j and K_k, of T_n.

(c) x_{n+1} is on the edge common to several elements K_i of T_n.

The fourth possibility corresponds to x_{n+1} being identical to one of the existing mesh points; this case is rejected as the points given are assumed to be distinct.

Using element(s) K_i, the set S of elements of T_n is created by a tree search, such that:

- x_{n+1} is internal to the circumsphere associated with the elements of S.

The triangulation T_{n+1} is constructed in the same way as in case a) above:

- Include the elements of T_n, not included in S, in T_{n+1} ,
- Remove the elements of S and remesh this set by joining point x_{n+1} to the external faces of S.

Once all points of the initial set have been introduced by this method, the covering-up of the initial hexahedron by tetrahedra is obtained, this mesh is called the "box" mesh (definition 11.2). This mesh covers the enclosing hexahedron and includes the points given, the eight points of the box and a few additional points[8], if necessary, as vertices (figure 11.9). In figure 11.10 we show both the mesh deduced from the box mesh by removing all tetrahedra with a vertex not included in the set of points of the domain, and the surface mesh of the object, i.e., the mesh used as data by the algorithm.

---

[8]In practice, this trick allows a significant improvement of the time required for the algorithm.

11.5. THE THREE-DIMENSIONAL CASE

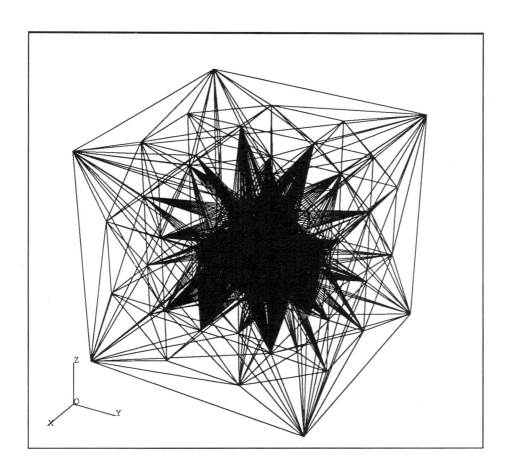

Figure 11.9: *"Box"* mesh.

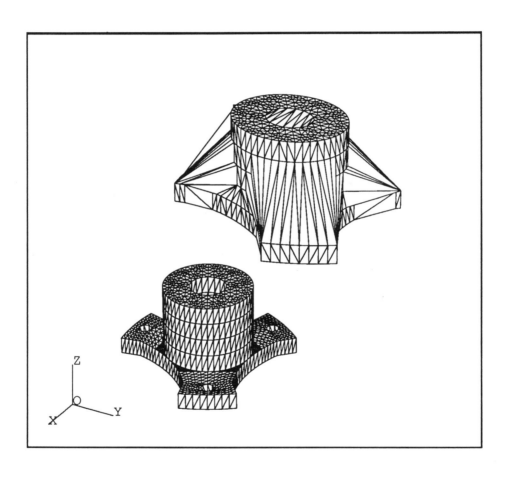

Figure 11.10: *Mesh of the envelope and surface mesh.*

11.5. THE THREE-DIMENSIONAL CASE

To create the boundary mesh using definition 11.3, we saw that (cf. section 11.4) it is sufficient to flag the elements of the box mesh by taking their location with respect to the boundary into account, and then to remove all the exterior elements, in 2 dimensions; in 3 dimensions, this process is generally not possible as the box mesh **does not contain**, in most cases, all the boundary faces in the list of its faces (known as the "hedgehog" effect [Coulomb-1987], [George et al. 1989a]) and, consequently, the notions of interior and exterior are not defined.

Some indications for solving this problem are proposed in the next section.

11.5.2 Regeneration of the boundary and suppression of the exterior

We propose a method to regenerate the boundary in the box mesh and refer to [Hermeline-1980], [Baker-1989] and [Perronnet-1988b] in which a different approach is investigated[9]. The mesh (of the box) does not contain all the triangular faces of the data. To overcome this difficulty, the problem is split into two parts:

1. Search for the missing edges and regenerate them, one by one,

2. After being assured of the existence of all the edges of the given boundary, search for all missing faces and regenerate them, one by one.

The process is as follows:

1. Missing edge case:

   For such an edge the set of tetrahedra for which the intersection with the associated segment is not empty, is defined. This segment is defined by the two endpoints of the edge under consideration. Only two situations occur:

   (a) The segment cuts element faces;
   (b) The segment cuts element edges.

   The set of elements connected with this segment is now modified locally in order to:

---

[9]There are, a priori, three approaches to solve this boundary problem: firstly, additional points can be created on this boundary to ensure that the box mesh includes the resulting discretization exactly; secondly, following the same principle, a post-treatment is then applied to remove these additional points; lastly, the mesh is modified in the neighbourhood of the boundary, this is the method discussed here.

- create the desired edge;
- suppress an undesirable edge.

This method for considering a missing edge is discussed in detail in [George et al. 1989a] and [George et al. 1989b]; the result, applied to all the missing edges, makes the creation of a mesh which includes all the boundary edges in the list of its edges possible. The proposed solution is based on local mesh modifications. For example, in two dimensions, the swapping of the diagonals of the quadrilateral formed by two adjacent triangles is a way of either creating a desired edge, or removing an undesirable edge (figure 11.11). Similarly, in three dimensions, it is possible to create a face by the removal of an edge or to create an edge by the removal of a face (figure 11.12). More elaborate transformations are obviously required, described in the two references mentioned above.

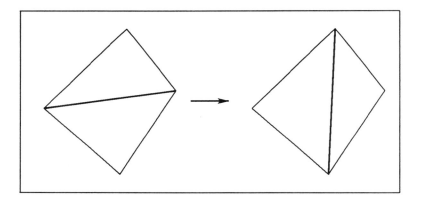

Figure 11.11: *Swapping of diagonals.*

2. Missing face case:

   Starting from the preceding result, the geometric situation with respect to missing faces is examined; the triangle associated with these types of faces is intersected by one or several edges. There is a demonstration of a method, in [George et al. 1989a] and [George et al. 1989b] to:

   - remove an undesirable edge and,
   - create a desired face.

11.5. THE THREE-DIMENSIONAL CASE

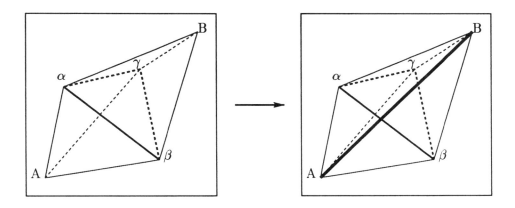

Figure 11.12: *Face-edge or edge-face.*

Applied to the above mesh, the method is suitable to create a new box mesh which fits the triangulation of the given boundary exactly (following definition 11.1).

A consequence of the latter property is that the notions of interior and exterior are now clearly defined. Consequently, the exterior elements of the object under consideration can be removed to obtain the boundary mesh.

Remark 11.5 : When regenerating the boundary, some internal points might be created to solve the problem locally. □

Remark 11.6 : The boundary mesh is no longer strictly of type *Delaunay* because some edges and faces have been created by local modification of the box mesh. This mesh is called a *constrained* mesh. □

Definition 11.4 : A mesh is called *constrained* if it must include a pre-specified collection of edges and faces in its respective lists of edges and faces. □

It is obvious that in the Voronoï approach, the notion of an edge or face to be preserved is a constraint, as the basic process consists only of a point insertion process which ignores this edge or face notion. This constraint is not present for other types of methods (for example those described in chapter 10).

11.5.3 Final mesh, element quality

The boundary mesh resulting from the previous step is not convenient for computing. It is necessary, as in the two-dimensional case, to modify it in order to obtain good quality tetrahedra, i.e., such that $Q = \frac{\rho}{h}$ (ρ is the radius of the circumsphere and h is the element diameter) is optimal[10]. To attain this result two techniques are recommended, based on:

- Taking only the volumes of the tetrahedra into account, with the aim of creating internal points, does not appear to be a suitable solution in the three-dimensional case.

- Internal point creation: the evaluation of the elements in the mesh by calculating their quality (i.e. their degeneracy), their volume and the consistency between the edge lengths and the local h of their vertices, is used to decide whether they are correct or not. If an element is judged incorrect for one or the other of these criteria, the optimal point associated with its smallest face is created. If this point is inside the element or inside one of its neighbours, it is inserted; if not the next tetrahedron is considered.

- Local mesh modifications: this operation consists of rewriting the elements adjacent by a face, an edge, or sharing a vertex [Talon-1987b].

A coupling of the last two operations appears to be very efficient.

The insertion of a point is done using the point insertion process mentioned above. Nevertheless, we should notice that in the case of a constrained mesh, this process must be adapted. Other strategies for internal point creation suitable in all cases are discussed in section 11.6. (see also [Baker-1989]).

Remark 11.7 : Similar to the two-dimensional case, the triangulation obtained is evidently dependent on the number and the relative position of the points discretizing the boundary. By specifying the latter, a variable density of elements can be obtained in certain regions. □

Remark 11.8 : The resulting mesh can be smoothed. □

This is the most general mesh generator we can define. Its applications are numerous, for example, it can be used for a large range of purposes by modifying different options (see for example [Boissonnat-1984] for an application in robotics).

---

[10]With respect to the result sought.

11.5. THE THREE-DIMENSIONAL CASE

11.5.4 Data description

As data, this mesh generator requires the data of a discretization of the boundary of the domain of interest. The boundary is formed by a mesh of triangular faces.

In addition some internal points, edges or faces can be specified in such a way that they will be included in the final triangulation.

In short, the data required by this mesh generator can be summarized as follows:

$$\text{Mesh of the boundary} + \text{additional constraints} \rightarrow \text{Mesh generator}$$

or

$$\mathcal{P} \rightarrow \text{Mesh generator}$$

in the particular case where the mesh of the envelope of the set is desired.

The face, edge and boundary point references of the created mesh are derived from those of the surface mesh provided as data. References not dealt with in this way are set to 0. The mesh generator will associate a sub-domain number, specified by the user, with all elements in the mesh.

11.5.5 Application examples

Module GHS3D[11] corresponds to this method and is used to illustrate this method.

Figure 11.13 presents two application examples of which the first corresponds to the mesh of a part of an homocinetic junction, consisting of 4468 tetrahedra and 1406 vertices; the second example is the mesh of a crank, including 3364 tetrahedra and 399 vertices.

This method lends itself well to the mesh creation of object exteriors (one can think of the domain defined by the body of an aircraft and a surface modelling infinity, an example of which was shown in the case of advancing front methods and is shown, for the present method, in [George et al. 1989b])

---

[11] GHS3D: Module developed at INRIA.

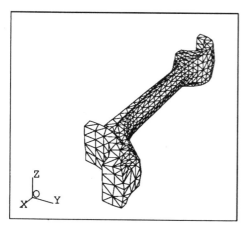 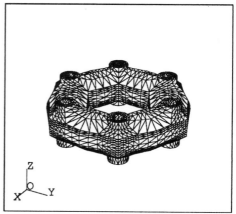

Figure 11.13: *Examples of final meshes (GHS3D)*.

11.6 Extensions

The following points will be discussed briefly in this section:

- A formal description of the insertion point process in a given mesh,

- The definition of a control space for governing the internal point creation phase in the boundary mesh,

- The utilization of this space to find the internal points,

- The general problem of transcribing a method of geometrical nature onto the computer.

11.6.1 Generalization of the point insertion process

The point insertion process described above (section 11.2) is elegant from a theoretical point of view and makes the construction of a triangulation with given vertices possible. Its application to the construction of a mesh generator satisfying some given objectives leads us to solve, on one hand (sections 11.4.2 and 11.5.2), the problem corresponding to the absence of specified edges or faces in the box mesh and, on the other hand (sections 11.4.3 and 11.5.3), to find an appropriate strategy for the internal point creation. To facilitate these operations, we first give a more general formulation of such a process.

11.6. EXTENSIONS

Denote (using the notation introduced in 11.2):

1. T_i a mesh containing the first i points of set $\{x_k\}$,

2. C_r a criterion (or a set of criteria),

3. C_o a constraint (or a set of constraints).

Supposing that mesh T_0 is given such that it contains all the points specified, then meshes T_i will be constructed step by step, by isolating the set S of elements in T_i such that:

($H1$) S satisfies criterion C_r with respect to point x_{i+1}
($H2$) S satisfies constraint C_o

Under the **sole** condition that it is possible to remesh S by taking point x_{i+1} into account (let S' be the set of the elements of this local remeshing), then the new mesh, T_{i+1}, is obtained by:

$$T_{i+1} = \{T_i - S\} \cup \{S'\}$$

An appropriate choice of C_r assures the validity of the point insertion process; for example, one can consider a criterion as simple as:

$C_r = \{S$ satisfies C_r for $x_{i+1} \Leftrightarrow$ its external faces are visible from $x_{i+1}\}$

or, in other words:

$C_r = \{S$ satisfies C_r w.r.t. point $x_{i+1} \Leftrightarrow S$ is star-shaped w.r.t. $x_{i+1}\}$

If the following definition of the criterion holds:

$C_r = \{S$ satisfies C_r with respect to point x_{i+1}

$\Leftrightarrow x_{i+1}$ is inside the circumcircle of every element of $S\}$

and if no constraint is imposed, the case described in section 11.2 is retrieved. It is therefore clear that a judicious choice of the couple $\{C_r, C_o\}$ can be used to obtain a triangulation enjoying interesting properties, in particular, at the level of the boundary regeneration phase and concerning an a priori control of the quality of the elements created ([George et al. 1989b]).

11.6.2 Control space

As for the advancing front approach (section 10.4.2), a control space is introduced, defined formally as follows:

Definition 10.2 :
(Δ, H) is a **control space** for mesh T of domain Ω if:

- $\Omega \subseteq \Delta$ where Δ is a covering-up enclosing Ω

- With each $P \in \Delta$ a functional $H(P, d)$ is associated, where d is a direction of sphere S^2 (of circle S^1 in dimension 2):

$$H(P, d) : \Delta \times S^2 \to R$$

□

Functional H, lying on the covering-up Δ, makes the specification of properties to be satisfied or criteria to be verified by the elements of the desired mesh possible. In addition, it provides a global knowledge of the domain to be meshed.

From a geometrical point of view, Δ is an arbitrary partitioning, for example, of one of the following types:

a) a partitioning of the type "quadtree" in dimension 2, or "octree" in dimension 3 (see section 8.3);

b) a regular partitioning (of finite difference type);

c) a pre-existing mesh (for instance the mesh T);

d) a pre-existing mesh specified by the user.

In addition to this covering-up feature, Δ includes, via the functional H, global information about the physical nature of the problem. The associated values can be used to evaluate if mesh T satisfies the functional at all points P (see the section below).

To obtain H, one of the following approaches can be taken:

- compute the local h associated with the points specified, and derive H by a generalized interpolation, (this analysis is purely geometrical in the sense that it relies on the geometrical properties of the data: size of the faces of the surface, etc.);

11.6. EXTENSIONS

- define, for each element of the covering-up Δ, the value of H manually (for example, specify the desired stepsize in the space - *isotropic* control - or along a direction - *anisotropic* control -);

- after an initial computation on a mesh of Ω, define H from the associated solution (or from its gradient, etc.), or from an error evaluation, ...;

- specify H manually by giving its value for each element of the covering-up constructed for this purpose (this is a control space of type d) seen above);

- in case a) above, the size of the boxes can be used to encode the value of H, meaning that the partitioning has been created so that the one or other property holds.

11.6.3 Control and internal points

From a given mesh (a boundary mesh if the internal point creation phase is considered, or a mesh which needs to be adapted to a physical solution), (Δ, H) is used to obtain a new mesh verifying the properties contained (encoded) in this control space.

Generally, the points are created in such a way that:

$$\| M - M' \| = H(M, \frac{M' - M}{\| M - M' \|})$$

for all points M', different from M, belonging to a same element K.

The quality of the elements is measured according to functional H (the usual definition $Q = \frac{\rho}{h}$, used before, corresponds to an isotropic and purely geometrical functional). In this generalization, the quality is defined by:

$$Q(K) = \rho/lmax$$

where:

$$lmax = 0.5 \max_{\text{edge AB}} \left(\frac{l_{AB}}{H\left(A, \frac{B-A}{\|A-B\|}\right)} + \frac{l_{BA}}{H\left(B, \frac{A-B}{\|B-A\|}\right)} \right)$$

with A and B the two endpoints of the edge under consideration, $l_{AB} = \| B - A \|$ and ρ the radius of the largest ball $B(P, \rho)$ inscribed in element K, with respect to H.

Remark 11.9 : This definition relies on the control of the elements via their edges; of course, one can choose other types of control. □

Remark 11.10 : The internal points can be constructed by another method while the associated connectivity is constructed by the present mesh generator. In this case, the control bears on the point creation method and not on the mesh generator. □

11.6.4 Computational geometry

To underline one of the difficulties encountered in the implementation of this mesh generation method (and more generally for all methods), we consider the following geometrical problem:

Let C be a given circle and P a point in the plane:

- is P inside C ?

- is P located on C ?

- is P exterior to C ?

It is easy to respond to these questions on paper by solving a simple equation (or a system of equations). It is less clear from a computational point of view on account of the rounds-off errors due to the accuracy of computers. In particular, it is easy to decide if P is *clearly* inside or outside C, but it is more difficult if P is close to C.

The first solution we can think of is to introduce a tolerance parameter ε; in this case, the problem is to choose a pertinent value: in practice, and whatever this choice is, the problem is not solved but only carried forward. The consequences of a bad response can produce not only a result which is partly erroneous, but a wrong result (in the present method, consider the effect of a bad decision when searching for the points which are inside the circumcircle of a triangle). It is therefore necessary to find a **consistant** description of the geometrical properties employed to obtain a reliable program.

Possible solutions for solving these types of problems are discussed in [George,Hermeline-1989]. More generally, the reader is referred to papers concerning "Computational geometry", which investigate this subject, amongst others (for example to [Preparata,Shamos-1985]).

Chapter 12

Creation of surface meshes

12.1 Introduction

The creation of surface meshes in the R^3 space proves useful in the two following situations:

- for numerical simulations on domains modelled by surfaces;
- for the creation of the data necessary for advancing front or Voronoï type volumetric mesh generators which require a surface mesh of the three-dimensional domain under consideration.

Excluding simple cases which consist, from a practical point of view, of creating a surface mesh by transformation of a two-dimensional mesh, we encounter the multiblock method, which deals with more complex geometries. It is natural to try to generalize the advancing front and Voronoï methods in $2D$ to surfaces; the problem is then to adapt them, by extending the purely $2D$ concepts (notion of circles, distances, etc.) to $2D^{\frac{1}{2}}$. This leads us, assuming that a suitable mesh generation method is available, to investigate the problem of defining a surface. This problem is at least as complex as that corresponding to the mesh creation. The surface description techniques in the C.A.D.[1] software packages seem the best solution for defining the geometry of surface domains.

There are numerous ways of presenting *curve and surface description* methods; each author having, for historical reasons, introduced his own notation and developed the subject for his specific application. Consequently, we do not claim to deal with the subject exhaustively within the

---
[1]Computer Aided Design: in fact, it corresponds to one of the many meanings of the word.

framework of this study. We will only try to point out some ideas so that the reader can become more familiar with the principal ideas of C.A.D. For a comprehensive study, the reader should consult the following references: [Coons-1967], [De Boor-1972], [De Casteljau-1985] and [Bézier-1986] who present some basic methods, [Thompson-1985] who gives a brief catalogue of the main curve approximation methods, and [Farin-1987] and [Farin-1988] who present a complete overview of the different curve and surface description techniques. For a more theoretical approach, [Fiorot, Jeanin-1989] and [Risler-1989] investigate approximations based respectively on rational functions, splines and Bezier patches. Obviously, these references constitute a small survey of the literature on this topic.

Concerning the creation of surface meshes, there is no readily available comprehensive reference to our knowledge, although some C.A.D. software includes surface mesh generators. It appears that a surface mesh, based on its description, is rather the result of a series of elementary operations (particular meshes) than the result of automatic processes of general applications. Nevertheless, we will try to give some directions regarding this problem.

First of all, the methods discussed in the previous chapters will be reconsidered, showing their applications to certain classes of geometries under certain (relatively restrictive) conditions. While trying to introduce an automatic method, we see that we are immediately up against the problem related to the definition of surfaces: the basic techniques of C.A.D. are then introduced. State of the art concerning mesh generation is outlined with special attention paid to the problems posed and the strong dependence between the surface description methods and the way they are meshed.

12.2 Transport-mapping method

It is possible to create quadrilateral elements by applying a space translation to a pre-meshed line (figure 12.1, left-hand side). This construction, a particular application of the method described in chapter 5, section 5.4, is only available when the surface can be defined in this manner.

A more general approach is a method based on a $2D$ mesh and a projection function to generate the surface mesh corresponding to the $2D$ input (conceptually, this method is similar to the previous one). The connectivity of the resulting mesh is the same as the input mesh. The point positions are computed with the help of the function specified.

It is clear that this type of construction only applies to surfaces which are sufficiently smooth with respect to a planar surface, for which a description in terms of a function is available (this holds for very particular geometries but is generally unrealistic).

12.3 Multiblock method

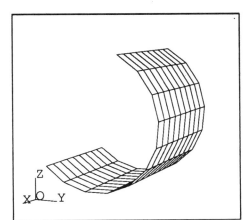
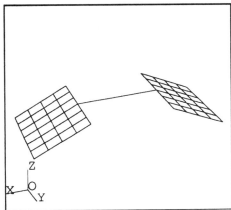

Figure 12.1: *Mesh of simple surfaces.*

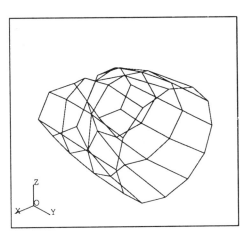
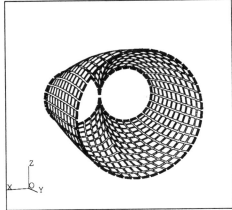

Figure 12.2: *Coarse mesh and resulting partitioning.*

The multiblock method (chapter 9) corresponds to the treatment of the problem by considering a coarse description of the domain to be meshed in terms of blocks of a triangular and a quadrilateral nature (figure 12.1, right-hand side and figure 12.2). In this case, the limitation of the preceding method is present: the surface must not be too deformed (the multiblock technique is based on a "mapping" (chapter 6), or a solution of a P.D.E. system (chapter 7), so that the limits of these two methods are clearly present). The real geometry of the domain is approximated by the data, i.e., by a coarse mesh. Consequently, all the information must be contained in this mesh. It will be shown later that this technique is a particular case of the general form of a polynomial approximation of a surface in each coarse element of the input mesh.

12.4 Automatic methods

One can try to extend the two-dimensional advancing front or Voronoï methods (cf. chapters 10 and 11) to surfaces: the triangles thus created are now defined in space. The problem is to define the space in which the quantities entering the computation are evaluated: the distance between two points, the relative location of a point with respect to a "circle", the definition of such a circle, etc. For example, the construction in chapter 11 (Voronoï method):

$$T_{i+1} = \{T_i - \mathcal{S}\} \cup \{\mathcal{S}'\}$$

is effective in practice if sets $\mathcal{S}$ and $\mathcal{S}'$ (cf. same chapter) can be defined properly.

In addition, if only the boundary points of the domain are given, the following questions concerning the internal point creation arise:

When is it necessary to create a point inside an element?

Where must this point be located?

This leads us to the surface description methods and, therefore, to C.A.D. techniques.

12.5 C.A.D. approach

The majority of C.A.D. packages allow the user to define a surface using the following tree-structure:

12.5. C.A.D. APPROACH

- Definition of points;

- Definition of curves;

- Definition of elementary surfaces;

- Definition of the surface under consideration with the help of the above items.

12.5.1 Points

Several types of points are to be considered: the control points which are used to define some higher order entities (curves and surfaces), the points of the curves and surfaces (which are not necessarily identical to the previous ones) and, lastly, the points created by the mesh generator on the surfaces (i.e. the element vertices of the mesh).

A point is either given explicitly or is the result of a computation (intersection of two curves, for example).

Furthermore, the points given can be present in the surface approximation or else merely serve as supports for information. In this case, they will not exist in this approximation but are used to define the set of points to be created on the surface.

12.5.2 Curves

The curves are created from points and relatively complex functions to ensure certain continuity properties (in particular at the junction of two curves). Two types of construction can be distinguished:

- the curve is defined by points and passes through them;

- the curve is defined by points but does not necessarily pass through them;

- the curve is defined by points and additional constraints (directional derivatives, etc.).

Some popular forms of approximate curve representations are now discussed; only polynomial approximations[2] are considered.

Let:

---

[2] Approximations exist which cannot be included in this formulation, see for example [Fiorot,Jeannin-1989].

n be an integer such that $n \geq 1$,

$V_0, V_1, ..., V_n$, be $n+1$ points in the space[3]

I be the interval $[a, b]$ where a and b are two real numbers such that $a < b$,

t be a parameter whose values $t_i, i = 0, n$, with $a = t_0 \leq t_1 \leq ... \leq t_n = b$, correspond to points $V_0, V_1, ..., V_n$;

r be a set of $n-1$ integers $r_1, ..., r_{n-1}$ such that $0 \leq r_i \leq n$.

The problem is to construct a piecewise polynomial function of t, of degree n and of class C^{r_i-1} in t_i (therefore at point V_i)[4]. The result is written in a form which is simple to implement on computer:

$$C(t) = [\mathcal{T}][\mathcal{M}][\mathcal{P}]$$

where:

$C(t)$ is the curve representation

$$[\mathcal{T}] = [t^n \quad t^{n-1} \quad ... \quad t \quad 1]$$

are the basis polynomials of the representation (a line vector)

$[\mathcal{M}]$ is a matrix of coefficients of dimension $n+1 \times n+1$.

$$[\mathcal{P}] =^t [V_0 \quad V_1 \quad .... \quad V_n]$$

is the control (column) vector, i.e., the set of values on which the approximation is based.

In particular, $[\mathcal{P}]$ assumes the following two forms:

$$[\mathcal{P}] =^t [P_0 \quad P_1 \quad .... \quad P_n]$$
$$[\mathcal{P}] =^t [P_0 \quad P_1 \quad .... \quad P_{\frac{n+1}{2}} \quad \dot{P}_0 \quad \dot{P}_1 \quad .... \quad \dot{P}_{\frac{n+1}{2}}]$$

---

[3] V_i denotes either a point in space: $V_i = P_i$, or a value associated with a point in space: for example $V_i = \dot{P}_i$ the derivative at point P_i.

[4] $r_i \geq 1$ denotes one constraint: at point P_i, so for value t_i, the function must be of class $r_i - 1$. For example, to obtain a continuity at point P_i, we set $r_i = 1$; to ensure a continuity at the tangent at point P_i, we set $r_i = 2$.

12.5. C.A.D. APPROACH

when the control only involves points and when it also involves derivatives.

Remark 12.1 : The choice of interval I and the repartitioning of the t_i is not indifferent. Formally, there is always an affine transformation making the mapping from any interval $[a,b]$ to an interval $[0,1]$ possible, and consequently, the latter is often selected. Although it is not a priori necessary, a uniform repartitioning of t_i is often assumed as it induces simpler computations. □

In practice, two classes of methods can be distinguished:

M_1 the curve is defined by values V_i, given in a unique definition;

M_2 the curve is defined as a set of sub-curves, each of them going through a sub-set of values taken from the given set; in this case, the management of the joining of two portions of the approximation is crucial.

Approximation of a curve with a unique definition

The following three curve approximation forms are presented, within the present framework, for arbitrary n :

1 the Lagrange form;

2 the Hermite form;

3 the Bezier form.

1 **Lagrange form**:

The function:

$$C(t) = \sum_{i=0}^{n} \phi_i(t) P_i \text{ where } \phi_i(t) = \prod_{l=0}^{n} \frac{t - t_l}{t_i - t_l}$$

with $l \neq i$

defines a curve, called the Lagrange interpolate, governed by the $n+1$ specified points which *passes through* these points: in fact, as

$$\phi_i(t_j) = \delta_{ij}$$

where δ_{ij} is the Kronecker delta, we have

$$C(t_i) = P_i$$

True for any n, this form can also be written in a recursive manner, as a polynomial of degree 1, which corresponds to the case $n = 1$, associated with $[t_i, t_{i+1}]$:

$$C_i^1(t) = \frac{t_{i+1} - t}{t_{i+1} - t_i} P_i + \frac{t - t_i}{t_{i+1} - t_i} P_{i+1}; \quad i = 0, ..., n-1$$

From this relation, the following is derived:

$$C_0^n(t) = \frac{t_n - t}{t_n - t_0} C_0^{n-1}(t) + \frac{t - t_0}{t_n - t_0} C_1^{n-1}$$

which defines the approximation $C(t)$ of the curve. One can verify that $C(t_i) = C_0^n(t_i) = P_i$, obvious for t_0 and t_n and easily extended to the other values t_i.

This recurrence is generalized in:

$$C_i^r(t) = \frac{t_{i+r} - t}{t_{i+r} - t_i} C_i^{r-1}(t) + \frac{t - t_i}{t_{i+r} - t_i} C_{i+1}^{r-1}(t); i = 0, ..., n-r \quad r = 1, ..., n$$

called the Aitkens algorithm.

In practice, the above expressions are simply two different forms of the same relation; they can be expressed in the simple matrix form introduced previously, as follows:

$$C(t) = [\mathcal{T}][\mathcal{M}][\mathcal{P}]$$

with:

$$[\mathcal{T}] = [t^n \quad t^{n-1} \quad ... \quad t \quad 1]$$
$$[\mathcal{P}] =^t [P_0 \quad P_1 \quad .... \quad P_n]$$

and $[\mathcal{M}]$ a $n+1 \times n+1$ coefficient matrix.

In the case where parameter t is chosen as: $t_i - t_{i-1} = \frac{1}{n}$, i.e., the sequence t_i is uniform, matrix $[\mathcal{M}]$ is written as follows:

$$\begin{pmatrix} -1 & 1 \\ 1 & 0 \end{pmatrix}$$

for $n = 1$ (assuming $[a, b] = [0, 1]$) and

$$\begin{pmatrix} 2 & -4 & 2 \\ -3 & 4 & -1 \\ 1 & 0 & 0 \end{pmatrix}$$

12.5. C.A.D. APPROACH

for $n = 2$ with the same assumptions. In the general case (cf. remark 12.1) we have, respectively:

$$\begin{pmatrix} \frac{1}{t_0-t_1} & \frac{1}{t_1-t_0} \\ \frac{-t_1}{t_0-t_1} & \frac{-t_0}{t_1-t_0} \end{pmatrix}$$

and

$$\begin{pmatrix} \frac{1}{(t_0-t_1)(t_0-t_2)} & \frac{1}{(t_1-t_0)(t_1-t_2)} & \frac{1}{(t_2-t_0)(t_2-t_1)} \\ \frac{-t_1-t_2}{(t_0-t_1)(t_0-t_2)} & \frac{-t_0-t_2}{(t_1-t_0)(t_1-t_2)} & \frac{-t_0-t_1}{(t_2-t_0)(t_2-t_1)} \\ \frac{t_1 t_2}{(t_0-t_1)(t_0-t_2)} & \frac{t_0 t_2}{(t_1-t_0)(t_1-t_2)} & \frac{t_0 t_1}{(t_2-t_0)(t_2-t_1)} \end{pmatrix}$$

2 Hermite form:

In addition to the input data required in the previous case, we specify:

$\dot{P}_0, \dot{P}_1, ..., \dot{P}_n$, the values of the $n+1$ derivatives at the preceding points,

Then the function:

$$C(t) = \sum_{i=0}^{n} \tilde{\phi}_i(t) P_i + \sum_{i=0}^{n} \varphi_i(t) \dot{P}_i$$

where

$$\tilde{\phi}_i(t) = \left\{ 1 - 2\dot{\phi}_i(t_i)(t - t_i) \right\} \phi_i(t)^2$$

and

$$\varphi_i(t) = (t - t_i) \phi_i(t)^2$$

defines a curve which is governed by the $n+1$ given points and their tangents. This curve, called the Hermite interpolate, is such that:

$$C(t_i) = P_i \quad \text{and} \quad \dot{C}(t_i) = \dot{P}_i$$

as can be shown.

This approximation can be written in the same way as above (using the same assumptions and notation):

$$C(t) = [\mathcal{T}][\mathcal{M}][\mathcal{P}]$$

with:

$$[\mathcal{T}] = \begin{bmatrix} t^n & t^{n-1} & \cdots & t & 1 \end{bmatrix}$$

$$[\mathcal{P}] =^t [P_0 \quad P_1 \quad .... \quad P_{\frac{n+1}{2}} \quad \dot{P}_0 \quad \dot{P}_1 \quad .... \quad \dot{P}_{\frac{n+1}{2}}]$$

and $[\mathcal{M}]$ a $(n+1) \times (n+1)$ coefficient matrix.

When $n = 3$, we have:

$$[\mathcal{T}] = [t^3 \quad t^2 \quad t \quad 1]$$

$$[\mathcal{P}] =^t [P_0 \quad P_1 \quad \dot{P}_0 \quad \dot{P}_1]$$

$$[\mathcal{M}] = \begin{pmatrix} 2 & -2 & 1 & 1 \\ -3 & 3 & -2 & -1 \\ 0 & 0 & 1 & 0 \\ 1 & 0 & 0 & 0 \end{pmatrix}$$

which is known as the cubic Hermite interpolation or the cubic Coons basis.

With the values P_i, $\dot{P}_i$ and the second derivatives in P_i, an interpolation of degree 5 can be constructed.

On the other hand, the following property[5] holds:

$$\dot{C}(t) = [\dot{\mathcal{T}}][\mathcal{M}][\mathcal{P}] = [nt^{n-1} \quad (n-1)t^{n-2} \quad .... \quad 1 \quad 0][\mathcal{M}][\mathcal{P}]$$

3 **Bezier curves** : An algebraic form of this approximation uses the Bernstein polynomials and, for arbitrary n, gives:

$$C(t) = \sum_{i=0}^{n} C_i^n t^i (1-t)^{n-i} P_i$$

with $C_i^n = \frac{n!}{(n-i)!i!}$

The De Casteljau algorithm:

$$P_i^r(t) = (1-t)P_i^{r-1}(t) + tP_{i+1}^{r-1}(t)$$

for $r = 1, n$, $i = 0, n - r$ and $P_i^0(t) = P_i$ control point number i

---

[5] General property of polynomial approximations.

12.5. C.A.D. APPROACH

also allows the creation of the Bezier curve of degree n governed by $n+1$ points:

$$C(t) = \sum_{i=0}^{n} \{P_0^n(t) P_i\}$$

Any Bezier curve can be written in the matrix form already introduced. For example, for $n = 3$, the general expression or the De Casteljau matrix $[\mathcal{M}]$ is given by:

$$[\mathcal{M}] = \begin{pmatrix} -1 & 3 & -3 & 1 \\ 3 & -6 & 3 & 0 \\ -3 & 3 & 0 & 0 \\ 1 & 0 & 0 & 0 \end{pmatrix}$$

with

$$[\mathcal{T}] = [t^3 \quad t^2 \quad t \quad 1]$$
$$[\mathcal{P}] = {}^t[P_0 \quad P_1 \quad P_2 \quad P_3]$$

Remark 12.2 : Starting with 4 points (retaining this case) and the Hermite form, this corresponds to the expression of the derivatives using 2 consecutive points. □

For large n, the above approximations either leads to oscillations or to prohibitive computations or even to impossible computations (there are terms of type t^n which, when $n = 100$ for example, lead to an overflow). Consequently, to benefit from the properties of these approximations and to avoid their disadvantages, the curves will be approximated by a set of sub-curves, each of them being defined using a representation of degree p, where $p < n$ is selected to be relatively small ($p = 1, 2, 3, 4, 5$, for example). The main problem turns to the choice of the representation in such a way that the joining between two portions of curves enjoys suitable properties (class C^0, C^1, C^2, etc).

Approximation of a curve by parts

Before giving some solutions to this problem, let us construct two simple curve representation examples. It corresponds to constructing a polynomial function approximating a curve enjoying certain properties. Consider:

— **Example 1** —

- n arbitrary, $p = 1$;

- $P_0, P_1, ..., P_n$ $n + 1$ points in the space and P_i, P_{i+1} 2 consecutive points in this series;

- $I = [t_i, t_{i+1}]$ mapped to $[0, 1]$ with t the parameter assumed to be evenly distributed;

- the curve passes through points P_i and P_{i+1} and its tangent at these points is prescribed to the value $\dot{P}_i$ ($\dot{P}_{i+1}$ respectively).

A polynomial of degree 3 satisfying 4 constraints is therefore sought. Let

$$C_i(t) = a_i + b_i t + c_i t^2 + d_i t^3 \text{ be this polynomial,}$$

where $C_i(0) = P_i$, $C_i(1) = P_{i+1}$, $\dot{C}_i(0) = \dot{P}_i$ and $\dot{C}_i(1) = \dot{P}_{i+1}$, then the following system holds:

$a_i = P_i$
$a_i + b_i + c_i + d_i = P_{i+1}$
$b_i = \dot{P}_i$
$b_i + 2c_i + 3d_i = \dot{P}_{i+1}$

It is evident that:

$$C_i(t) = [\mathcal{T}][\mathcal{M}][\mathcal{P}_i]$$

with

$$[\mathcal{T}] = [t^3 \quad t^2 \quad t \quad 1]$$
$$[\mathcal{P}_i] =^t [P_i \quad P_{i+1} \quad \dot{P}_i \quad \dot{P}_{i+1}]$$

and

$$[\mathcal{M}] = \begin{pmatrix} 2 & -2 & 1 & 1 \\ -3 & 3 & -2 & -1 \\ 0 & 0 & 1 & 0 \\ 1 & 0 & 0 & 0 \end{pmatrix}$$

which is nothing other than the cubic Hermite form with the following properties:

- $C(t_i) = C_{i-1}(t_i) = C_i(t_i) = P_i$ alias $C_{i-1}(1)$ and $C_i(0)$;

- $\dot{C}(t_i) = \dot{C}_{i-1}(t_i) = \dot{C}_i(t_i) = \dot{P}_i$.

12.5. C.A.D. APPROACH

The curve passes through the points and is of class C^1.

— **Example 2** —

- n is indeterminate, $p = 3$;
- $P_0, P_1, ..., P_n$ $n+1$ points in the space and $P_{i-1}, P_i, P_{i+1}, P_{i+2}$ 4 consecutive points of this series;
- $I = [t_i, t_{i+1}]$ mapped to $[0, 1]$ and t the evenly distributed parameter;
- $r_i = 2, r_i + 1 = 2$, i.e., the curve passes through points P_i and P_{i+1} and its tangent at these points is fixed to the value $\dot{P}_i = 0.5(P_{i+1} - P_{i-1})$ (resp. $\dot{P}_{i+1} = 0.5(P_{i+2} - P_i)$).

As above, a polynomial of degree 3 satisfying 4 constraints is sought: $C_i(0) = P_i$, $C_i(1) = P_{i+1}$, $\dot{C}_i(0) = 0.5(P_{i+1} - P_{i-1})$ and $\dot{C}_i(1) = 0.5(P_{i+2} - P_i)$.

This implies that:

$a_i = P_i$
$a_i + b_i + c_i + d_i = P_{i+1}$
$2b_i = P_{i+1} - P_i$
$2b_i + 4c_i + 6d_i = P_{i+2} - P_i$

It is easy to see that the above representation holds:

$$C_i(t) = [T][M][\mathcal{P}_i]$$

with

$$[T] = [t^3 \quad t^2 \quad t \quad 1]$$

$$[\mathcal{P}_i] = {}^t[P_{i-1} \quad P_i \quad P_{i+1} \quad P_{i+2}]$$

and

$$[M] = 0.5 \times \begin{pmatrix} -1 & 3 & -3 & 1 \\ 2 & -5 & 4 & -1 \\ -1 & 0 & 1 & 0 \\ 0 & 2 & 0 & 0 \end{pmatrix}$$

which is nothing other than the Catmull-Rom form with the following properties:

- $C(t_i) = C_{i-1}(t_i) = C_i(t_i) = P_i$ alias $C_{i-1}(1)$ and $C_i(0)$;

- $\dot{C}(t_i) = \dot{C}_{i-1}(t_i) = \dot{C}_i(t_i) = \dot{P}_{i+1} - \dot{P}_{i-1};$

The curve passes through points $P_i, i = 1, n-1$ using, in addition, points P_0 and P_n, is of class C^1.

Remark 12.3 : The construction method illustrated by these examples is suitable for the construction of different polynomial approximations varying by the data and the specified constraints. □

In the following, the principal forms for the approximation of curves will be given. To obtain these functions, either the method proposed in the two above examples can be followed, or a general form can be studied directly by fixing the value of p. The same result is evidently found in both of these cases.

- **Lagrange form:**

  For any n, fix $p = 1$, then the Lagrange form can be written for each interval $[t_i, t_{i+1}]$ as:

  $$C_i(t) = \frac{t_{i+1} - t}{t_{i+1} - t_i} P_i + \frac{t - t_i}{t_{i+1} - t_i} P_{i+1}, \quad t \in I$$

  The degree of this approximation to the curve is 1, the only property present is: $C_i(t_i) = C_{i-1}(t_{i+1})$, i.e., the parts are joined in C^0. In this case, the curve is approximated by function $C(t)$ which is a polyline composed of the different $C_i(t)$.

  To obtain a higher degree approximation for each part, it suffices to choose $p > 1$, for example for $p = 2$ the following matrix $[\mathcal{M}]$ is obtained:

  $$\begin{pmatrix} 2 & -4 & 2 \\ -3 & 4 & -1 \\ 1 & 0 & 0 \end{pmatrix}$$

  The joining between the $C_i(t)$ is also in C^0: nothing better can be obtained by this method.

- **Hermite form:**

  Fix p and use the following function on each interval $[t_i, t_{i+1}]$:

  $$C_i(t) = \sum_{i=0}^{p} \tilde{\phi}_i(t) P_i + \sum_{i=0}^{p} \varphi_i(t) \dot{P}_i$$

12.5. C.A.D. APPROACH

with the definitions specified above. In this way, the curve is governed by the $n+1$ given points and by their tangents. For example, for $p = 3$ and for each $C_i(t)$, we return to the matrix representation of the cubic Coons basis:

$$C_i(t) = [T][\mathcal{M}][\mathcal{P}_i]$$

with:

$$[T] = [t^3 \quad t^2 \quad t \quad 1]$$
$$[\mathcal{P}_i] =^t [P_i \quad P_{i+1} \quad \dot{P}_i \quad \dot{P}_{i+1}]$$

Clearly, in this case, the interpolation passes through the points and the joining between two parts is in C^1. If, in addition, the second derivatives are given and $p = 5$, the resulting joining is in C^2.

- **Bezier cubic basis**: starting from 4 points (we reconsider this example ($p = 3$) in that which follows), we construct an approximation of the curve $C(t)$ by a series of curves $C_i(t)$ defined locally as Bezier curves of degree 3. Then:

$$[\mathcal{M}] = \begin{pmatrix} -1 & 3 & -3 & 1 \\ 3 & -6 & 3 & 0 \\ -3 & 3 & 0 & 0 \\ 1 & 0 & 0 & 0 \end{pmatrix}$$

with

$$[T] = [t^3 \quad t^2 \quad t \quad 1]$$
$$[\mathcal{P}_i] =^t [P_i \quad P_{i+1} \quad P_{i+2} \quad P_{i+3}]$$

- **Catmull-Rom polynomial basis**: to achieve a C^1 continuity at the endpoints of the sub-curves defined by 4 points and, in this way, to overcome the problem present in the above method, we define the derivative by considering the values on both sides of the point under consideration. The above relation becomes:

$$[\mathcal{P}_i] =^t [P_{i-1} \quad P_i \quad P_{i+1} \quad P_{i+2}]$$

$$[\mathcal{M}] = 0.5 \times \begin{pmatrix} -1 & 3 & -3 & 1 \\ 2 & -5 & 4 & -1 \\ -1 & 0 & 1 & 0 \\ 0 & 2 & 0 & 0 \end{pmatrix}$$

Each sub-curve is defined from point P_i to point P_{i+1} but uses the values at points P_{i-1} and P_{i+2} for its construction (as for example 2).

- **Cardinal-Spline basis**: a generalization of the previous representation, this form involves a control parameter, α, which measures the amplitude of the vectors tangent to the control points.

$$[\mathcal{P}] = {}^t [P_{i-1} \quad P_i \quad P_{i+1} \quad P_{i+2}]$$

$$[\mathcal{M}] = \begin{pmatrix} -\alpha & 2-\alpha & \alpha-2 & \alpha \\ 2\alpha & \alpha-3 & 3-2\alpha & \alpha \\ -\alpha & 0 & \alpha & 0 \\ 0 & 1 & 0 & 0 \end{pmatrix}$$

When $\alpha = 0.5$, the Catmull-Rom form is retrieved.

- **B-Spline basis**: to obtain a C^2 continuity, the constraint imposing the curve to pass through the specified points is removed, giving:

$$[\mathcal{P}] = {}^t [P_{i-1} \quad P_i \quad P_{i+1} \quad P_{i+2}]$$

$$[\mathcal{M}] = \frac{1}{6} \times \begin{pmatrix} -1 & 3 & -3 & 1 \\ 3 & -6 & 3 & 0 \\ -3 & 0 & 3 & 0 \\ 1 & 4 & 1 & 0 \end{pmatrix}$$

The general form of this representation, for any k, is given by:

$$C_i(t) = \sum_{i=0}^{k} B_k(t) P_i$$

where $B_k(t)$ is defined (for $k = 3$) as being $B_{0,3}(t)$. This term is constructed by recurrence as follows:

$$B_{i,k}(t) = \frac{t - t_i}{t_{i+k} - t_i} B_{i,k-1}(t) + \frac{t_{i+k+1} - t}{t_{i+k+1} - t_{i+1}} B_{i+1,k-1}(t)$$

with $B_{i,0}(t) = 1$ if $t \in [t_i, t_{i+1}[$ and 0 otherwise.

Assuming $k = 3$ and $t_i = i$ for simplicity (for an evenly partitioned interval), the above matrix is still obtained. In fact:

$$B_{0,3}(t) = \frac{1}{3} \{t B_{0,2}(t) + (4 - t) B_{1,2}(t)\}$$

12.5. C.A.D. APPROACH

is computed from:

$$B_{0,2}(t) = \frac{1}{2}\{tB_{0,1}(t) + (3-t)B_{1,1}(t)\}$$

$$B_{1,2}(t) = \frac{1}{2}\{(t-1)B_{1,1}(t) + (4-t)B_{2,1}(t)\}$$

which are computed from:

$$B_{0,1}(t) = \{tB_{0,0}(t) + (2-t)B_{1,0}(t)\}$$
$$B_{1,1}(t) = \{(t-1)B_{1,0}(t) + (3-t)B_{2,0}(t)\}$$
$$B_{2,1}(t) = \{(t-2)B_{2,0}(t) + (4-t)B_{3,0}(t)\}$$

using the value of $B_{i,0}$, we obtain successively:

$$B_3(t) = \frac{t^3}{6} \quad t \in [0,1[$$

$$B_3(t) = \frac{1}{6}(-3t^3 + 12t^2 - 12t + 4) \quad t \in [1,2[$$

$$B_3(t) = \frac{1}{6}(3t^3 - 24t^2 + 60t - 44) \quad t \in [2,3[$$

$$B_3(t) = \frac{1}{6}(-t^3 + 12t^2 - 48t + 64) \quad t \in [3,4[$$

to define $C_i(t)$ on $[0,1]$, it suffices to make a change of variable: $t \to t-i$ in the four expressions of $B_3(t)$ to obtain the $C_i(t)$ form which can be written in the same matrix form given above, as will be verified as an exercise.

By introducing a parameter α, this form can be generalized, as before, by controlling the shape of the curve.

- **Beta-Spline basis** : by introducing two parameters β_1 (the bias) and β_2 (the tension) in a B-Spline form, the behaviour of the curve is controlled by moving it towards the control points.

$$[\mathcal{P}] = {}^t[P_{i-1} \quad P_i \quad P_{i+1} \quad P_{i+2}]$$

$$[\mathcal{M}] = \frac{1}{\Delta}\begin{pmatrix} -2\beta_1^3 & 2(\beta_2 + \beta_1^3 + \beta_1^2 + \beta_1) & -2(\beta_2 + \beta_1^2 + \beta_1 + 1) & 2 \\ 6\beta_1^3 & -3(\beta_2 + 2\beta_1^3 + 2\beta_1^2) & 3(\beta_2 + 2\beta_1^2) & 0 \\ -6\beta_1^3 & 6(\beta_1^3 - \beta_1) & 6\beta_1 & 0 \\ 2\beta_1^3 & \beta_2 + 4(\beta_1^2 + \beta_1) & 2 & 0 \end{pmatrix}$$

with $\Delta = \beta_2 + 2\beta_1^3 + 4\beta_1^2 + 4\beta_1 + 2$. For $\beta_1 = 1$ and $\beta_2 = 0$, the classic B-Spline form is found.

For $\beta_1 = 1$, we have the **Beta2-Spline basis**.

Remark 12.4 : It is possible to derive one form from the other by simple matrix operations. □

Remark 12.5 : As some curves cannot be represented correctly with a polynomial basis, other approximations, for example using rational Splines, can be constructed. □

To sum up, the definition of the representation of a curve can be performed:

- either in a global way (method M_1) by using, for example, the Lagrange, Hermite or Bezier form;

- or by parts (method M_2) by using a form for each sub-curve which induces the desired properties at the level of their joining points.

The representation passes through the control points, or not. It allows, for each value of parameter t, to evaluate the position of the corresponding point.

12.5.3 Elementary surfaces

The different methods to construct a curve can be extended to a surface. Similarly, the required functions are chosen in such a way that continuity properties are present.

Let u be a second parameter, then we assume that the control points depend on this parameter:

$$P_i \to P_i(u)$$

The general form discussed in the case of curves applies to surfaces. More precisely, a patch is defined by $\{(t, u) \in [a_1, b_1] \times [a_2, b_2]\}$. Then we have:

$$C(t) = [\mathcal{T}][\mathcal{M}][\mathcal{P}]$$

which becomes

$$C(t, u) = [\mathcal{T}][\mathcal{M}][\mathcal{P}(u)]$$

with

$$P(u) = [\mathcal{U}][\mathcal{M}][\mathcal{P}]$$

12.5. C.A.D. APPROACH

which gives (denoting $S(t,u)$ to be the result):

$$S(t,u) = [\mathcal{T}][\mathcal{M}][\mathcal{P}_{ij}]^t[\mathcal{M}]^t[\mathcal{U}]$$

with:

[$\mathcal{M}$] the matrix associated with a curve representation (cf. section 12.5.2.)
[$\mathcal{U}$] the equivalent in u to [$\mathcal{T}$], i.e., the associated basis polynomials and
[$\mathcal{P}_{ij}$] the control points, where i denotes the dependence with respect to parameter t, and j that with respect to parameter u, [$\mathcal{P}_{ij}$] is a $n+1 \times n+1$ matrix constructed on the control points.

A formal and more general form of $S(t,u)$ is:

$$S(t,u) = \sum_{i=0}^{n}\sum_{j=0}^{m} \left\{ b_{ij} t^i u^j \right\}$$

where b_{ij} depends on the method selected (n and m may be different). In the case where a Bezier form is used, $S(t,u)$ can be expressed in terms of the Bernstein polynomials:

$$B_i^n(t) = C_i^n t^i (1-t)^{n-i}$$

by

$$S(t,u) = \sum_{i=0}^{n}\sum_{j=0}^{m} \left\{ B_i^n(t) B_j^m(t) P_{ij} \right\}$$

This form is known as the representation of a surface by Bezier patches.

Basically, this method only allows the definition of quadrilateral patches. But, for flexibility, it is necessary to be able to use triangles, which will be introduced, either directly, or as a degenerate case of the preceding patches[6].

Denote:

- Δ a triangle in R^2;

- P_0, P_1, P_2 its three vertices;

- r, s, t the barycentric coordinates of point P in R^2 with respect to the three P_i above;

---

[6] Historically, triangles were introduced first.

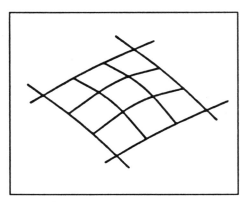 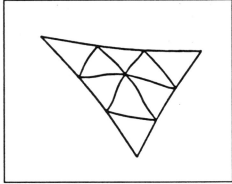

Figure 12.3: *Quadrilateral and triangular patches.*

Define, for n a fixed integer and Δ a given triangle, the following polynomials:

$$B_{i,j,k}^n(r,s,t) = \frac{n!}{i!j!k!} r^i s^j t^k \text{ for } i+j+k = n$$

A triangular face is then defined as the mapping from Δ onto R^3 as follows:

$$B(r,s,t) = \sum_{i+j+k=n} B_{i,j,k}^n(r,s,t) P_{i,j,k}$$

where $P_{i,j,k}$ is a series of points in R^3 forming a triangular network ($i+j+k = n$), which are the control points of the representation.

The definition of a surface representation is done, as will be shown below, using these elementary surfaces (a simple surface can be described in terms of a single elementary surface).

The representation passes through the control points, or not. The location of the point corresponding to each couple of parameters, t and u, is calculated using this representation.

12.5.4 Global description of a surface

A surface is defined as a series of elementary surfaces. The joining of them ensures the continuity of the total surface. As elementary surfaces, we encounter the quadrilateral and the triangular patches seen above. For illustrating purposes, we assume that we have two additional types of elementary surfaces to our disposal:

12.5. C.A.D. APPROACH

- a patch "corroded" at one of its corners, and

- a patch with a "hole".

for which the region of interest is the zone remaining or its complementary (cf. figure 12.4). In practice, the two basis patches are merely particular cases of these extensions.

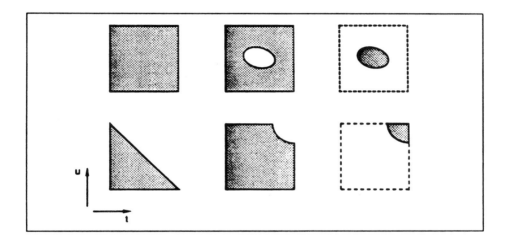

Figure 12.4: *Basis patches.*

To be more precise, we reconsider the example in the definition of Catmull-Rom with $p = 3$. The surface under consideration, assumed complex, is described by a series of patches which are constructed using the control points and the representation chosen. Each patch is joined to its neighbours with the properties present in this representation.

The surface is defined by a coarse grid[7], derived from the control points used in the case of the description for the Catmull-Rom method. Thus, to define a grid on the surface which includes q_1 divisions in one direction (t) and q_2 in the other (u), one has to provide $q_1 + 3$ series of $q_2 + 3$ control points. The endpoints only serve to define the shape of the surface by its boundary. Each patch is then determined by its four points and by the points of its neighbours (cf. figure 12.5), passing through the four points which define it and enjoying a continuous junction with its neighbouring patches as the points of the latter are taken into account in its definition.

---

[7]By definition 1.4.

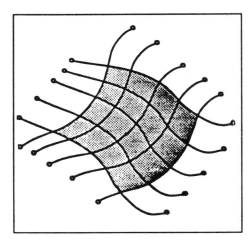 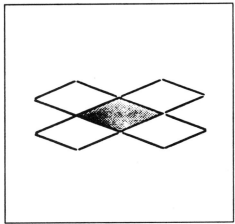

Figure 12.5: *Surface description by patches and one of these patches.*

12.6 Mesh generation

We assume that the surface under consideration is known by a series of elementary surfaces of the types described above. The global mesh can be seen as the union of the different elementary surface meshes. In order to obtain a valid mesh (with respect to definition 1.1), it is necessary to ensure the following property:

- any point (element vertices of the mesh) common to two patches must be defined in the same way in each patch which contains it;

This implies that the lines bordering each patch are meshed in the same way in all the patches containing them. Thus, the scheme below must be followed:

- For each patch, do:
    - For each boundary line, do:
        * if the line has already been meshed (when considering a patch processed previously), consider its discretization (its mesh),
        * else, mesh the line in such a way that its mesh is compatible with that of the lines already meshed.

12.6. MESH GENERATION

When all the lines forming the boundaries of the patches have been discretized, the mesh of all the patches is created. This mesh is a function of the nature of the patch.

1- If the patch is quadrilateral or triangular and if none of its boundary lines contains intermediary points, it is considered as an element in the mesh;

2- If the patch is quadrilateral or triangular and if all its boundary lines contain a given number of intermediary points compatible with a regular partitioning (cf. chapter 6), it is meshed by a suitable method;

3- If the patch is arbitrary, an automatic mesh generator is used.

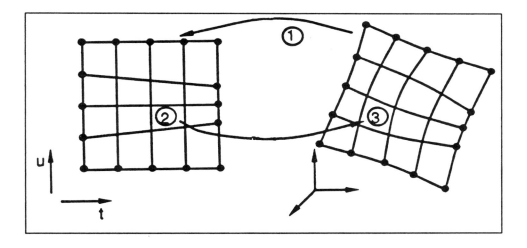

Figure 12.6: *Mesh of a patch.*

As case 1 is obvious, let us consider case 2, while considering the Catmull-Rom example and the case concerning a patch split according to two parameters n_1 and n_2. The mesh of this patch can be written in the following algorithmic form:

Do (step 1):

- Do for $i = 0$ and $i = n_1$

  - Do for $j = 0, n_2$

    * consider the location in R^3 of node i, j (located on a boundary line previously meshed);
    * compute the value of the associated parameters t, u.

- Do for $j = 0$ and $j = n_2$

  - Do for $i = 0, n_1$

    * consider the location in R^3 of node i, j (cf. above note);
    * compute the value of the associated parameters.

Note that this step is not trivial!

Then (step 2):

Create the mesh in the space t, u of the unit square $[0, 1] \times [0, 1]$ as a function of its boundary discretization.

Lastly (step 3):

- Do for $i = 1, n_1 - 1$

  - Do for $j = 1, n_2 - 1$

    * definition of the connectivity: the vertices of the element created have the following couples as vertex numbers: (i, j), $(i+1, j)$, $(i+1, j+1)$ and $(i, j+1)$, each of which will have a global number associated with it (see for example chapter 9)
    * computation of the location of these vertices: evaluate t and u corresponding to i and j and find the location using the formula:

    $$S(t, u) = [T][\mathcal{M}][\mathcal{P}_{ij}]^t[\mathcal{M}]^t[\mathcal{U}]$$

    with $[\mathcal{P}_{ij}]$ the control matrix of the current patch.

  End do for j

End do for i

12.6. MESH GENERATION

This returns to applying a 2D method by considering the t, u parameter space.

In the case of an elementary triangular surface, a similar algorithm can give the same solution with a different representation for the evaluation of the nodes of the mesh.

Case 3 (automatic mesh) is currently far from being solved adequately, in general. Any 2D method can be implemented in the t, u space, the problem being then to know if the mapping in R^3 of the mesh points in the space of the parameters is valid and, more precisely, to know if:

- the elements created are close to the surface, i.e., the surface is well approximated ?

- the elements created are of good quality (cf. section 11.6.3. for example) ?

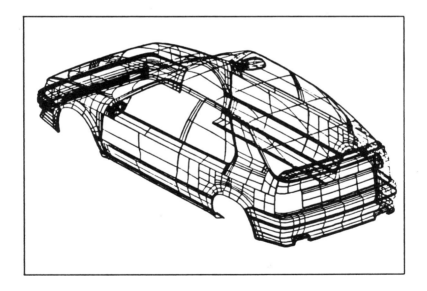

Figure 12.7: *The C.A.D. description of a R19 body (by Renault).*

Figure 12.7 shows, in the example of a R19 body (model with 3 doors), the set of patches used for its surface description. This representation was produced by the Euclid software (cf. chapter 14). Figure 12.8 presents the mesh corresponding to a description of the same type for the 5 door model.

212   CHAPTER 12. CREATION OF SURFACE MESHES

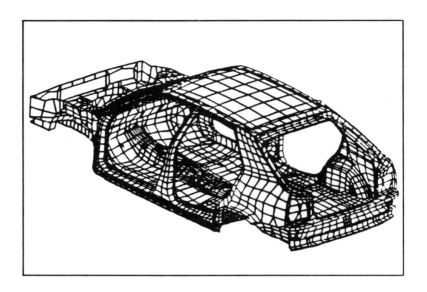

Figure 12.8: *The surface mesh of a R19 body (ibidem).*

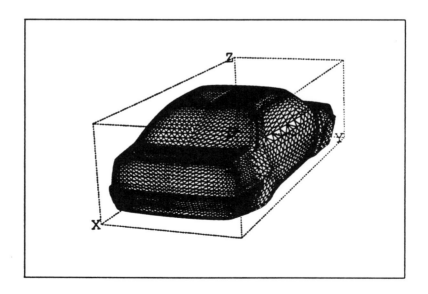

Figure 12.9: *The surface mesh of a 405 body (by Peugeot).*

12.6. MESH GENERATION

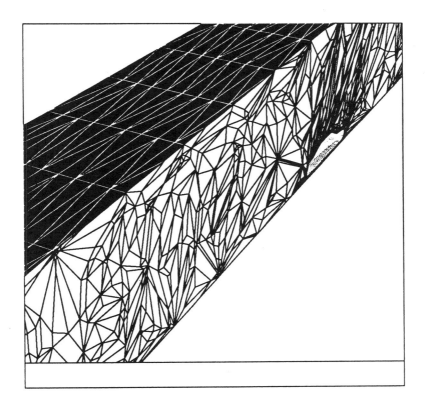

Figure 12.10: *Mesh of a 405. Research for the application of a volumetric mesh generator from the surface data of a car body.*

The mesh, produced by I-DEAS-Supertab (cf. chapter 14) on half of the body, contains 4466 nodes and 4555 shell elements. The mesh of this figure corresponds to the pasting together of this half-mesh with its symmetrical counterpart. Figure 12.9 shows the mesh of a 405 and figure 12.10 a cut through this $3D$ mesh created around the body (the latter mesh is output by mesh generator GHS3D in chapter 11)[8].

Remark 12.6 : The imperatives of C.A.D. (i.e. the description of the surfaces for graphic or manufacturing purposes) are not, strictly, identical to those of the finite element method. In particular, the details necessary in the first approach are not always useful in numerical simulation. □

Remark 12.7 : We are aware of having only outlined the subject, but we hope to have underlined the principal problems which arise. The generation of surface mesh could by itself be the object of a complete book. □

---

[8]We thank Peugeot and Renault for supplying the data of these two significant examples.

Chapter 13

Mesh transformations

13.1 Introduction

In this chapter, we assume that we have one or several meshes, obtained by one or other mesh generator at our disposal. The different mesh transformations possible are presented. The following aspects will be approached:

- The usual geometrical transformations which allow the creation of new meshes by the transformation of existing ones;

- A method for the fast and reliable pasting together of two meshes sharing a common region of lower dimension (a line or a point in $2D$; a surface, a line or a point in $3D$);

- Local and global refining techniques;

- An approach producing the regularization of the mesh elements as a function of different constraints;

- The list of mesh modifying tools and their application to the adaptation of the meshes with respect to the specific properties of the problems under consideration;

- Some notes about particular treatments of existing meshes in order to impose particular properties on them (internal constraints (internal points, lines or surfaces to respect), modification of attributes, generation of the nodes (if they are distinct from the vertices), renumbering of the nodes, etc.);

- Finally, some details concerning the graphic representation of meshes, their visualization and related problems, are given.

13.2 Usual geometrical modifications

For simplicity, we consider "P1" meshes (the element nodes are their only vertices).

Assuming that a mesh is given then a new mesh can be created by applying any usual geometrical transformation to it (symmetry, translation, rotation, anisotropic dilation or any type of transformation explicitly defined by its matrix).

In the case where the transformation corresponds to a *positive isometry*, it only changes the position of the mesh points; thus the connectivity and the numbering of the elements remain unchanged.

In the case where the transformation corresponds to a *negative isometry*, it acts on the position of the mesh points and, in our situation, on the numbering (or connectivity) of the elements, in such a way that the positive surface elements (in dimension 2), or the volumetric elements (in dimension 3), are created. To achieve this, a permutation of the list of the element vertices must be made (for example, a triangle with vertices $(1)(2)(3)$ is transformed into the triangle with vertices $(1)(3)(2)$).

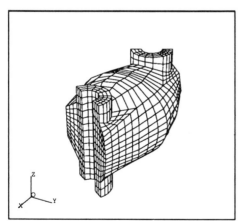 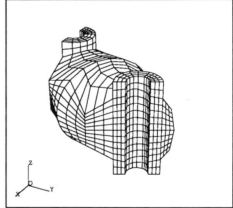

Figure 13.1: *Geometrical manipulation of a mesh (case of a rotation).*

The different transformations of interest are:

- Symmetries with respect to a line (in dimension 2) whose equation is given, or with respect to a plane (in dimension 3) defined in the same way;

13.2. USUAL GEOMETRICAL MODIFICATIONS

- Translations of a given vector;

- Rotations of a given angle about a point in dimension 2; about an axis and with respect to a centre of rotation in dimension 3 (figure 13.1);

- Isotropic or anisotropic dilations with respect to a given centre where the coefficients in 2 (3) directions are given.

In addition, any combination of these operators define a new transformation.

Formally speaking, and following the geometrical point of view, these operations can be summarized by a transformation of the type:

$$M \Rightarrow M' \text{ with } M' = Mat(M)$$

where Mat is a matrix.

To define these transformations explicitly, we assume that we are in a homogeneous coordinates system. Thus, in the two-dimensional case, the following definitions hold:

- for a **symmetry** with respect to line $Ax + By + C = 0$, the matrix Mat is written as:

$$Mat = \begin{pmatrix} 1 + A^2F & ABF & ACF \\ ABF & 1 + B^2F & BCF \\ 0 & 0 & 1 \end{pmatrix} \text{ with } F = \frac{-2}{A^2 + B^2}$$

- for a **translation** of vector $\vec{T} = (T_x, T_y)$, the matrix Mat is:

$$Mat = \begin{pmatrix} 1 & 0 & T_x \\ 0 & 1 & T_y \\ 0 & 0 & 1 \end{pmatrix}$$

- for a **rotation** of angle α about a point $P = (P_x, P_y)$, the matrix Mat is given by:

$$Mat = \begin{pmatrix} \cos\alpha & -\sin\alpha & P_x \\ \sin\alpha & \cos\alpha & P_y \\ 0 & 0 & 1 \end{pmatrix}$$

- in the case of a **dilation** with centre $C = (C_x, C_y)$ and with coefficients α_x, α_y, we have:

$$Mat = \begin{pmatrix} \alpha_x & 0 & C_x(1-\alpha_x) \\ 0 & \alpha_y & C_y(1-\alpha_y) \\ 0 & 0 & 1 \end{pmatrix}$$

For the three-dimensional case:

- for a **symmetry** with respect to plane $Ax + By + Cz + D = 0$, the matrix Mat is defined by:

$$Mat = \begin{pmatrix} 1+A^2F & ABF & ACF & ADF \\ ABF & 1+B^2F & BCF & BDF \\ ACF & BCF & 1+C^2F & CDF \\ 0 & 0 & 0 & 1 \end{pmatrix} \text{ with } F = \frac{-2}{A^2+B^2}.$$

- for a **translation** of vector $\vec{T} = (T_x, T_y, T_z)$, the matrix Mat is:

$$Mat = \begin{pmatrix} 1 & 0 & 0 & T_x \\ 0 & 1 & 0 & T_y \\ 0 & 0 & 1 & T_z \\ 0 & 0 & 0 & 1 \end{pmatrix}$$

- for a **rotation** of angle α, with axis $A = (A_x, A_y, A_z)$ defined about a point $P = (P_x, P_y, P_z)$, the matrix Mat is given by:

$$\begin{pmatrix} a^2+Sd-CS_2g & ab+Se-CS_2h & ac+Sf-CS_2i & P_x \\ -SC_2a+Cd+SS_2g & -SC_2b+Ce+SS_2h & -SC_2c+Cf+SS_2i & P_y \\ S_2a+C_2g & S_2b+C_2h & S_2c+C_2i & P_z \\ 0 & 0 & 0 & 1 \end{pmatrix}$$

with:

- If $A_x \neq 0$ then:
  $\phi = \arctan(\frac{A_y}{A_x})$, $\theta = \arctan(\frac{A_z}{\sqrt{A_x^2+A_y^2}})$ and $C = \cos\phi$,
  $S = \sin\phi$, $C_2 = \cos\theta$, $S_2 = \sin\theta$
- If $A_y \neq 0$ then:
  $\theta = \arctan(\frac{A_z}{\sqrt{A_x^2+A_y^2}})$ and $C = 0$, $S = 1$, $C_2 = \cos\theta$,
  $S_2 = \sin\theta$

13.2. USUAL GEOMETRICAL MODIFICATIONS

- Otherwise
$C = 1, S = 0$ and $C_2 = 0, S_2 = 1$

and:
$a = C_2 C$
$b = -C_2 S$
$c = S_2$
$d = \cos \alpha S + \sin \alpha S_2 C$
$e = \cos \alpha C - \sin \alpha S S_2$
$f = -\sin \alpha C_2$
$g = \sin \alpha S - \cos \alpha S_2 C$
$h = \sin \alpha C + \cos \alpha S S_2$
$i = \cos \alpha C_2$

It is possible to verify that this matrix is nothing other than the combination of the following matrices:

$M(-\phi, \vec{Z}) \circ M(-\theta, \vec{Y}) \circ M(\alpha, \vec{X}) \circ M(\theta, \vec{Y}) \circ M(\phi, \vec{Z})$ expression in which $M(angle, vector)$ denotes the matrix associated with the rotation of angle *angle* and axis *vector* (the centre of rotation is supposed to be at the origin; to return to the general situation, it corresponds to adding the translation which takes into account the point P above).

- in the case of a **dilation** with center $C = (C_x, C_y, C_z)$ and with coefficients $\alpha_x, \alpha_y, \alpha_z$, we have:

$$Mat = \begin{pmatrix} \alpha_x & 0 & 0 & C_x(1-\alpha_x) \\ 0 & \alpha_y & 0 & C_y(1-\alpha_y) \\ 0 & 0 & \alpha_z & C_z(1-\alpha_z) \\ 0 & 0 & 0 & 1 \end{pmatrix}$$

Remark 13.1 : In the case of a rotation, the definition of angles ϕ and θ by an arctangent gives a determination at modulo π. When programming such an operator, this ambiguity must be removed using the sinus and cosinus of the angles in such a way that they are well determined (in Fortran, the function $ATAN2$ allows this specification). □

Remark 13.2 : When applying a geometrical transformation of a mesh, it might be necessary to modify certain physical attributes: one sub-domain number of the initial mesh $\Rightarrow$ another number for the final mesh ... (see section 13.8). □

13.3 Pasting together of two meshes

The pasting process consists of constructing the mesh resulting from the juxtaposition of two initial meshes after identification of the common zones (points, edges and faces). To obtain an efficient process, a "haching" technique (h-coding technique, [Aho et al. 1983]) can be used, which locates the items common to the two meshes.

Remark 13.3 : The pasting region of two meshes must verify the conformity properties (definition 1.1). □

Remark 13.4 : The pasting together of two meshes, a very natural and frequent operation in mesh generation packages, can be seen in a very different way. In fact, a corollary of the conception methodology is that the pasting zone is formed by a set of two lines or two surfaces, and therefore by their constitutive items (points, edges and faces), a priori known, i.e., not sought in one way or another. In fact, the storage of this information in a mesh data structure is not trivial, consequently we must regenerate them by computation. This apparent weakness is compensated for, on one hand, by the efficiency of the pasting algorithms and, on the other hand, by the resulting modularity in the sense that the pasting formally becomes independent of the conception of the two meshes. □

The pasting is therefore based on the search of items common to two meshes. This operation consists of several phases:

- The determination of the *geometrically common* points: to find the list of the common points, a method based on a "h-coding technique" ([Aho et al. 1983]) is proposed.
  - The two initial meshes are enclosed in a quadrilateral "box" (hexahedral in $3D$) where the dimensions, $\Delta x, \Delta y,$ and Δz, are computed;
  - Let ϵ be a given tolerance, then the stepsize is given by
  $$\epsilon_\eta = \epsilon \times max(\Delta x, \Delta y, \Delta z)$$
  which is in fact the resolution of the algorithm;
  - The space is partitioned into blocks of size ϵ_η;
  - The points of meshes 1 and 2 are then defined with respect to these blocks.

The potential pasting points, in terms of the blocks introduced, are then sought:

13.3. PASTING TOGETHER OF TWO MESHES

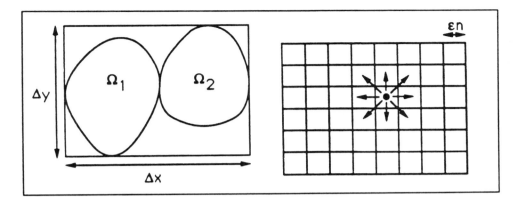

Figure 13.2: *Definition of the pasting blocks.*

- Consider a point belonging to mesh 1, and check if there exist points in mesh 2 contained inside the block enclosing this point or inside the adjacent blocks;

- If one or several candidates are found, it suffices to verify that: $sup|x_1^i - x_2^i| < \epsilon_\eta$ where x_1^i is the coordinate i of the point belonging to the first mesh and x_2^i that of the point belonging to the second mesh; in this case the two points are identical.

Remark 13.5 : Depending on the number of blocks (clearly connected to ϵ_η), and the number and position of the points, there may be *collapses*. This corresponds to the case where a block contains several points belonging to one or other mesh. The searching algorithm for the common points must take this situation into account. □

The list of the points common to the two meshes is obtained by this method. This technique is very efficient at the computer level.

- The determination of the *common* points: two points are called common if they are obtained by this process. Due to physical reasons, we refine this notion of coincidence. Thus, the following two options are taken into account:

  1. Option 1 : Purely geometrical coincidence: two points are common if they satisfy the above test.

  2. Option 2 : Physical coincidence: two points are common, in

this respect, if they satisfy the same test and, secondly, if their physical attributes (reference number) are identical.

Assuming this definition, it is possible to distinguish two points which are identical in terms of position, but different by the conditions to which they are subject (for example, a crack can be created in this way).

- The determination of the *common* edges: two edges are called common if their two endpoints are common by the above criteria; in the same way, the physical attribute of an edge is used to obtain the same distinction as above.

- The determination of the *common* faces: similarly, two faces are called common if their vertices are common; the physical number of the face is used as before.

It suffices to take the result of the previous analysis into account to obtain the mesh resulting from the pasting of two initial meshes. The points of this new mesh are composed of the points of the first mesh and those of the second mesh which are not common. To obtain the numbering of this second series of points, the common points retain the numbers they had in the first mesh and the others receive the first available free number, incremented by one for each point processed. The elements of the new mesh are, firstly, those of the first initial mesh and then those of the second mesh enumerated in sequence, the vertices of these elements numbered using the value found when constituting the list of the new point numbers.

Following this construction, neither the element nor the vertex numbering is optimal (except in particular situations). Consequently, it is advisable to proceed with a renumbering (see section 13.10).

13.4 Specified internal points, lines or faces

The presence of several materials in a domain or the need to assign some particular conditions to a point, edge, face or set of these items, may lead to the specification of a point, edge, face or set of these members inside a domain.

This type of property represents a constraint to be respected by the mesh of the domain, which must include all the imposed items in the list of its constitutive items (points, edges and faces).

13.4. SPECIFIED INTERNAL POINTS, LINES OR FACES

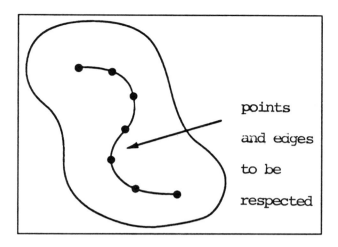

Figure 13.3: *Internal constraint.*

To arrive at this result, two approaches can be investigated. These two points of view will now be analyzed.

1. Incorporate the constraint in the analysis of the problem, in particular at the level of the definition of the primal sub-sets: the prescribed set is displayed explicitly and used in the definition of the primal sub-sets created during the analysis of the domain. Thus, for example, a prescribed line becomes a part of the contour of two primal sub-sets. The constraint is respected at the time they are meshed; after pasting them together, it is naturally present in the final mesh.

2. Use a mesh generation algorithm which incorporates the constraint: if this constraint involves the contour of the domain, we obtain the previous situation; if not, it must be considered inside the generator itself. There are two cases:

    - Case of an advancing front type mesh generator: if this approach only considers domains with boundaries possessing a single connected component (see remark 10.3), it falls under this case. The constraint is composed of a set of points, lines and faces E, which is artificially joined to contour C of the domain. This junction results in the creation of a new contour as follows:

    $$C = C + E_1 + E_2$$

where:

- E_1 is constraint E defined in the appropriate direction,
- E_2 is constraint E defined in the opposite direction,

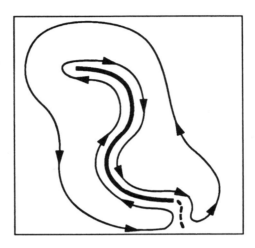

Figure 13.4: *Constrained contour (two-dimensional case)*.

This contour definition is done in such a way that the boundary only has one connected component and is described in the right direction (to determine the interior of the domain).

- Case of a Voronoï type mesh generator: the constraint is integrated in the mesh generator data:
  - If it corresponds to points: they will be included in the list of points, serving as data, and will be inserted into the mesh by the insertion process described in section 11.6.1 and, consequently, will be present in the final mesh.
  - If it corresponds to edges: these edges will be included in the list of edges, after the boundary edges, and must be present in the set of edges of the final mesh. This type of mesh with a constraint is not, in general, obtained as soon as the endpoints of the specified edges are inserted. A method is proposed in [George et al. 1989b] for the construction, both in dimension 2 and dimension 3, of a mesh containing prespecified edges, by local modification of a mesh containing the points corresponding to this specification.

13.4. SPECIFIED INTERNAL POINTS, LINES OR FACES

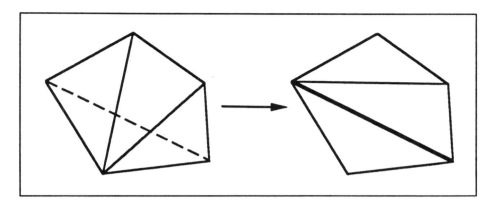

Figure 13.5: *Local modification (two-dimensional case)*.

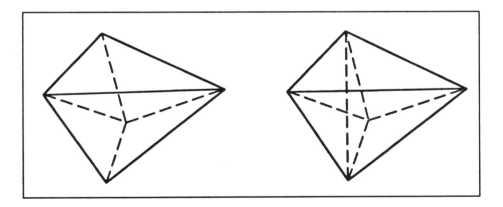

Figure 13.6: *Local modification (three-dimensional case)*.

- If it corresponds to faces: to our knowledge, this type of constraint is very rarely encountered in codes; the algorithm described in section 11.7. is one of the rare ones for this type of specification. The reader is referred to this section and to [George et al. 1989b] for a more thorough investigation consisting of a full description of a method ensuring this property.

13.5 Local or global refinements

13.5.1 Global point of view

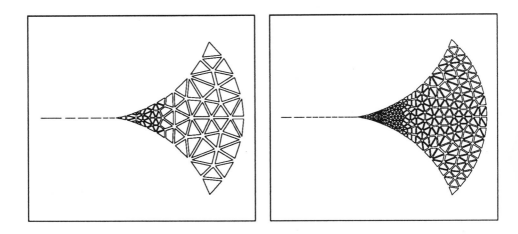

Figure 13.7: *Global refinement (N=2)*.

Assuming that a mesh and a splitting parameter, N, are given then N intermediary points are defined on each edge of the mesh. The canonical partitioning process, already described in chapter 6, is used to split all the elements into $(N+1)$ elements (segment case), $(N+1)^2$ elements (triangle and quadrilateral cases), or into $(N+1)^3$ elements (tetrahedral, pentahedral and hexahedral cases). The elements resulting from the splitting are of the same type as the element from which they are derived.

13.5.2 Local point of view

Several algorithms can be considered for the local mesh refinement in the neighbourhood of some of its vertices. Such a process can be based on:

13.5. LOCAL OR GLOBAL REFINEMENTS

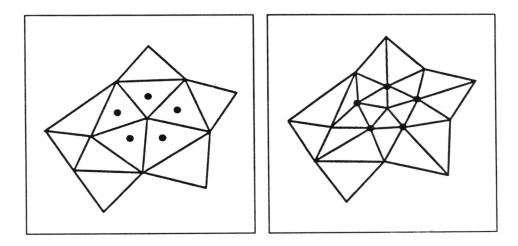

Figure 13.8: *Local transformation (1): addition of points.*

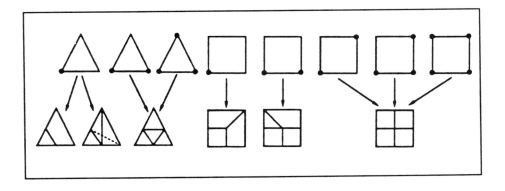

Figure 13.9: *Local transformation (2): refinement around some vertices.*

- A local remesh around some selected points. In the case when the meshes are composed of *simplices* (triangles in 2D and tetrahedra in 3D) a method of the type described for generators of type "Voronoï" is suitable, coupled with local modifications [Talon-1987a],[Talon-1987b].

- A partitioning [Bank et al. 1983] of the selected elements by bisection, trisection, etc. can be applied (see figure 13.9 for the local aspect and figures 13.10 and 13.11 for the application of this process); this method generates a partitioning of the neighbouring elements for conformity reasons. While it is easy to implement such a method in two dimensions, this type of local modification is more difficult to conceive and implement in three dimensions.

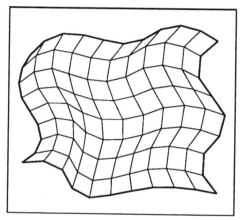
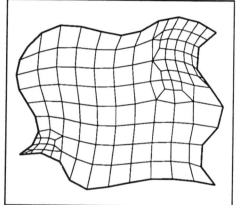

Figure 13.10: *Initial mesh and refinement in two zones.*

In particular, the necessity to create conformal meshes induces the splitting of elements near the element under consideration and, in this way, the refinement propagates (figures 13.10 and 13.11). To avoid this kind of constraint, adequate transitional finite elements can be defined.

- A deformation of elements; in this case, the connectivity (i.e. the junction between the vertices) is unaffected but the vertex positions are modified (cf. section 13.5.4.).

13.5.3 Type modification

13.5. LOCAL OR GLOBAL REFINEMENTS

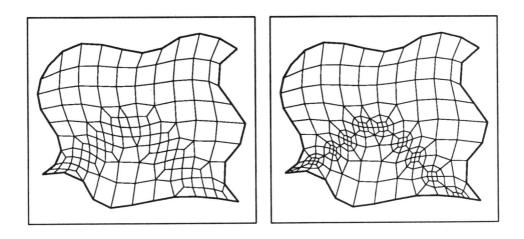

Figure 13.11: *Refinement along a line (states 1 and 2).*

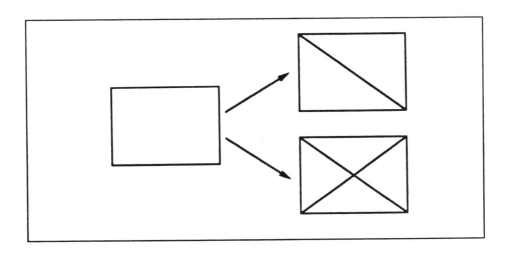

Figure 13.12: *Change of type (quadrilateral case).*

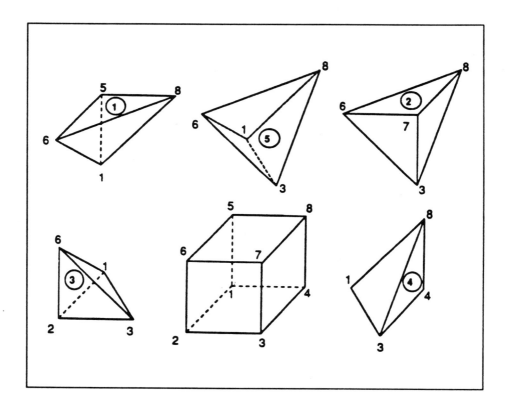

Figure 13.13: *Splitting a hexahedron into five tetrahedra.*

13.5. LOCAL OR GLOBAL REFINEMENTS

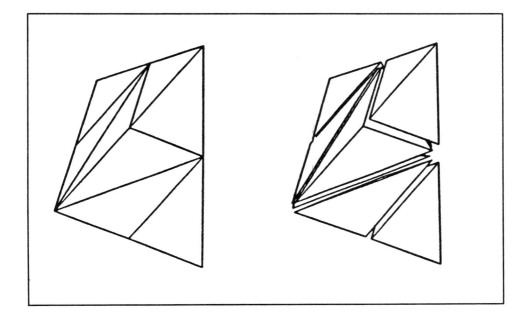

Figure 13.14: *Splitting of a tetrahedron at the midsides (doc. Vénéré).*

This operation, applied to quadrilaterals, creates two or four triangles; applied to pentahedra, it generates 3 tetrahedra; in the case of hexahedra, it produces 5 or 6 tetrahedra.

Inversely, the splitting of a triangle into quadrilaterals and that of a tetrahedron into hexahedra can be defined; generally, this type of operation leads to bad results.

A mesh composed of triangles in the plane can be modified in such a way that quadrilaterals are created by joining, two by two, the triangles adjacent by an edge. Some triangles may of course remain in the final mesh, as a function of the number of points present.

13.5.4 Deformations

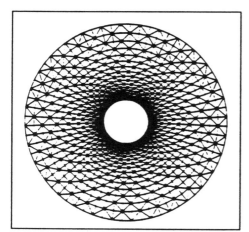 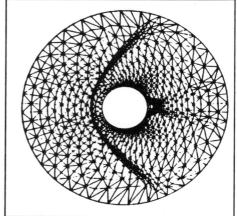

Figure 13.15: *Element deformations (initial and final states) (by Palmerio)*.

For a given connectivity, a mesh can be refined locally by deforming some selected elements (for example, those located in a region with a stiff gradient). This modification acts on the vertex positions of the elements dealt with. In practice, an iterative method is used which permits the easy control of the deformation process to avoid the overlapping, turning over, etc. of elements. Figure 13.15 shows the application of this technique for a simple geometrical situation [Palmerio-1988]. The solution computed with each of these meshes is displayed in figure 13.16; the interest of the deformation seems therefore obvious. Coupled with an enrichment, the deformation technique results in the creation, given an initial state, of a mesh adapted to the physical properties of the problem under consideration

13.5. LOCAL OR GLOBAL REFINEMENTS

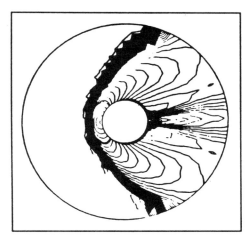
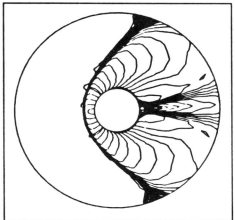

Figure 13.16: *Corresponding isomach lines (ibidem).*

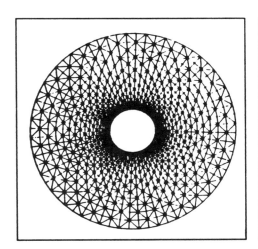
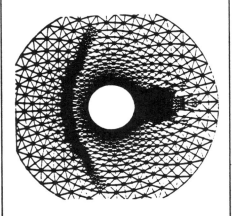

Figure 13.17: *Deformations and enrichment (initial state and mesh resulting after two enrichment steps) (ibidem).*

(concerning the adaptation, see also section 13.7). Figure 13.17 shows this type of example taken from [Palmerio-1987] in the case of the Euler equations.

13.6 Mesh regularization

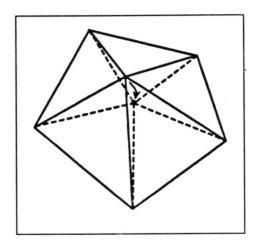

Figure 13.18: *Displacement of a point at the barycentre of its neighbours.*

Besides the weighted barycentre techniques to smooth a mesh, we have iterative methods which are based on a displacement of the internal mesh points coupled with a control of the resulting effect (for example in terms of quality):

- Barycentrage: The set of points neighbouring a given point P is determined; they correspond to the vertices of the elements having point P as vertex. Point P is then relocated to the weighted barycentre of its neighbours (figure 13.18 for the local effect and 13.19 for the global effect):

$$P = \frac{1}{n} \sum_{j=1}^{n} \alpha_j P_{k_j}$$

where n denotes the number of points connected to point P, P_{k_j} denote these points with numbers k_j, and α_j is the associated weight

13.6. MESH REGULARIZATION

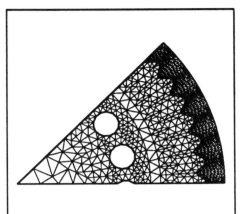
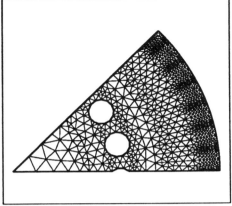

Figure 13.19: *Initial and smoothed mesh by barycentrage.*

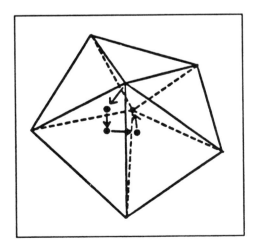

Figure 13.20: *Displacement of a point by controlled iterations.*

(with $\sum \alpha_j = 1$.). A variant consists of "relaxing" this expression, described by the following expression:

$$P^{m+1} = (1-\omega)P^m + \frac{\omega}{n} \sum_{j=1}^{n} \alpha_j P_{k_j}^m$$

with identical notation and ω a relaxation parameter. This operation may create points exterior to the domain for a non-convex domain Ω, and a control is therefore required.

- Iterative method: let P be a vertex of the mesh and Q be an associated measure (for example, the maximum of the quality of the elements with P as vertex), then an iterative process for the displacement of P along an arbitrary direction is defined, where the amplitude is initialized to some given value. The displacement of P is simulated and the associated value Q is computed by successive iterations; the way in which it varies directs the process (figure 13.20). In the non-convex case, the validity of the result must be verified at each iteration.

Remark 13.6 The regularization can be interpreted as a mesh optimization. As an application, a new mesh generation process can be formed which can be summarized as:

- The data of a random mesh (i.e. with a valid connectivity but where the point positions are arbitrary);

- The optimization of this mesh in order to obtain a correct mesh. □

13.7 Adaptation

The adaptation of meshes to the physical properties of the problems under consideration is a technique which is used more and more. It leads to the computation of better solutions as the computations are done on meshes composed of elements adapted to the physical problem under consideration. In this way, we deal with optimal meshes in terms of the size required (number of nodes), node positions and element properties (element shape, density, orthogonality, etc.).

A combination of regularization techniques and local splitting of elements seems to be a good approach for the construction of these types of adapted meshes.

In the first phase, a mesh of the domain is created by one of the methods proposed previously, after which an initial computation of the solution of

13.7. ADAPTATION

the problem is made. After a choice of a pertinent criterium involving the gradient of the solution, a derived field or an evaluation (generally a discrete evaluation) of the interpolation error, we detect the zones to adapt in the initial mesh (by refinement, derefinement, local remeshing, ..., see below) and a new, better adapted, mesh is generated; this is generally an iterative process.

Adaptation tools

- Repositioning of nodes or *r-method*: assuming an invariant connectivity (junction between nodes), the nodes are relocated either by weighted barycentrage defined by the position and the weight of their neighbours, or by deformation of some elements.

- Global or local enrichment or *h-method*: in this approach, the results mentioned in section 13.5 concerning the splitting of selected, or all the, elements, can be used; in the case where the mesh is composed of simplices, the point insertion process (section 11.2) can be employed to add points to an existing mesh.

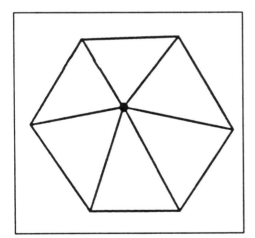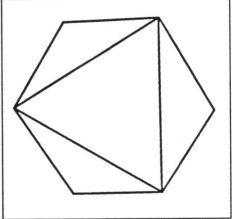

Figure 13.21: *Removal of a vertex.*

- Local disenrichment[1]: in the case of simplicial elements, an algorithm can be designed to remove one or several points. This operation modifies the elements locally in the neighbourhood of this point. The

---
[1]This technique can be considered as an example of the *h-method*.

basic idea is that if a vertex belongs to only three triangles (four tetrahedra), it can be removed: to belong to this case (figure 13.21) some adequate diagonal swappings (two-dimensional case), or some *edge-face* changes (section 11.5.2), are made (an optimization of the result is in general necessary in order to avoid the creation of flat elements).

- Nested meshes and multigrid techniques: this type of approach calls for a comprehensive and complex study; we thus refer to [Hackbush,Trottenberg-1982].

- Local remeshing[2] (case of meshes composed of simplices): in this case, we return to the configurations studied in the Voronoï methods.

From a formal point of view, the adaptation can no longer be described in a local fashion, but via a global approach. The next section aims to give a comprehensive overview of this topic.

Control space

In a general manner, the adaptation of a mesh can be seen as follows:

Existing mesh + Solution + Criteria $\Rightarrow$ adapted mesh

To express the criterium, or the set of criteria, it is convenient to introduce the notion of a *control space*, denoted by (Δ, H).

In the absence of a global definition of this space (Δ, H), it can be created from the mesh containing the points given as input with which an isotropic stepsize, computed from the relative positions of these points, or from the edges or faces to which the point under consideration belongs, is associated.

The control space is therefore, in this case, governed by the data, i.e., in the majority of cases, by the mesh of the contour of the domain to be meshed.

There are several choices available to obtain the stepsize h at each point P:

- from the lengths and respective directions of the edges with point P as endpoint;

---

[2]See note 1.

13.7. ADAPTATION

- from the surface and orientation of the faces with P as vertex.

Thanks to this control definition, the measure H is given at each point in the mesh, for example in the case of a tetrahedral element:

$$H(P,d) = \left(\sum_{i=1,4} b_i h(P_i, d)^p \right)^{\frac{1}{p}}$$

where the P_i are the vertices of tetrahedron T containing point P, b_i a weight associated with the P_i's, p a value which interacts with the refinement propagating from the boundary. Functional H is designed in the same way for other types of elements.

Thus $H(P,d)$ is a generalized continuous interpolation.

The control space (Δ, H) can be fully defined in order to prescribe certain properties to the final mesh of domain Ω.

Formally, (Δ, H) is an arbitrary covering-up enclosing Ω such that a functional H is defined on all its elements.

To elaborate this point, we can mention the following three types of partitionings Δ used to define the control space (Δ, H) (cf. figures 13.22, 13.23 and 13.24):

a) a partitioning of the type "quadtree" in $2D$, or "octree" in $3D$;

b) a regular partitioning;

c) a pre-existing mesh.

The definition of type c) applies to the case where an initial mesh of Ω is known with which the solution of the physical problem under consideration associates the values of functional H at all points. This couple (mesh of $\Omega + H(P,d)$) constitutes the space (Δ, H) from which the creation of a new mesh of Ω will be undertaken. The elements of this remeshing must satisfy the properties defined by (Δ, H); consequently, an adapted mesh is created.

The example shown in figure 13.25 corresponds to an application of these ideas. In this case [Vallet-1990] we have chosen:

- as space Δ: an initial mesh obtained by an arbitrary method. This mesh is created by considering only the geometrical aspect of the domain.

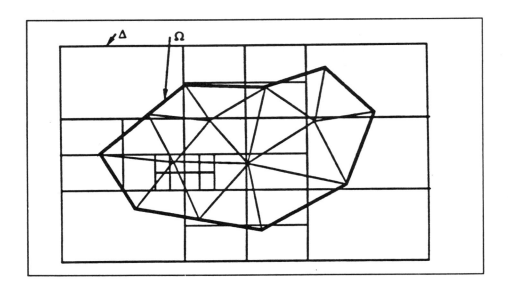

Figure 13.22: *Control space of type* **a**).

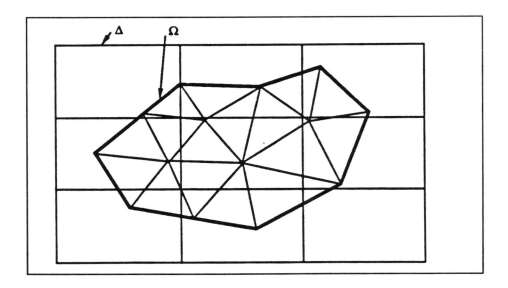

Figure 13.23: *Control space of type* **b**).

13.7. ADAPTATION

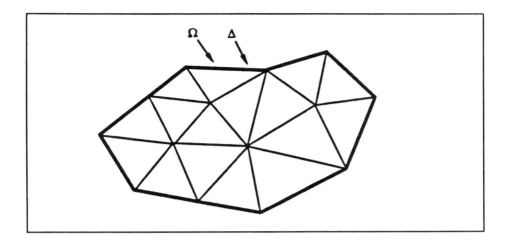

Figure 13.24: *Control space of type* **c**).

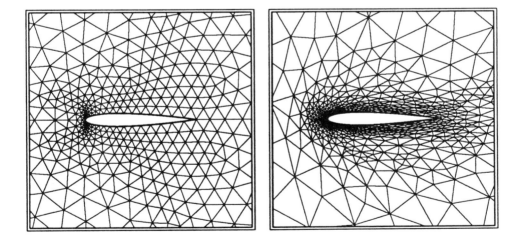

Figure 13.25: *Anisotropic mesh (by Vallet)*.

- as functional H: a solution of the Navier-Stokes problem (for which the mesh must be adapted) is computed on the initial triangulation. The velocity distribution (denoted by u) obtained is used to compute the following Hessian matrix in terms of the u norm:

$$\begin{pmatrix} \frac{\partial^2 \|u\|}{\partial x^2} & \frac{\partial^2 \|u\|}{\partial x \partial y} \\ \frac{\partial^2 \|u\|}{\partial x \partial y} & \frac{\partial^2 \|u\|}{\partial y^2} \end{pmatrix}$$

A metric is defined by this matrix. The new mesh is then created in such a way that this metric is satisfied at each point.

Remark 13.7 : Refer also to chapters 10 and 11, in which this notion of a control space was introduced in the case of the advancing front and Voronoï methods. □

Remark 13.8 : By modifying the degree of the polynomials used for the interpolation (and no longer the mesh), we return to the *p-methods*. □

13.8 Modifying the physical attributes

Used in numerous steps in the mesh generation process, the different changes of numbers are done either along with the geometrical change or explicitly. Within the frame of the proposed methodology, they are used to create geometrically equivalent meshes whose physical attributes are different from those of the initial mesh; thus the produced meshes can be used in a different context.

In this process, it is necessary to, on one hand, assign a sufficient quantity of numbers to the desired items in the initial mesh and, on the other hand, specify the way in which they must be modified. This correspondence (figure 13.26) is in fact strongly related to the type of transformations envisaged (see also section 3.4).

Remark 13.9 : In some mesh generation codes, this generation is a purely geometrical process. The physical attributes (reference and subdomain numbers) are only defined at an ulterior phase. In this case, a processor is to be created and the necessary information is usually enumerated by blocks. □

13.9. GENERATION OF NON-VERTEX NODES

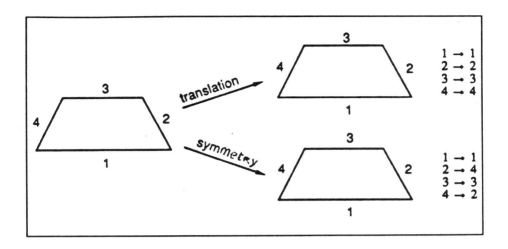

Figure 13.26: *Transfer of attributes and geometrical transformation.*

13.9 Generation of non-vertex nodes

When the finite elements used in the computation are not $P1$ elements (i.e. elements whose nodes are not only the vertices), it is necessary to add the non-vertex nodes (on edges, faces or inside elements) and, if desired, to indicate that the vertices do not coincide with the nodes (see figure 13.27).

As the local node numbering is predefined (cf. section 2.3.5.), a global number must be assigned to them.

A method facilitating this task consists of creating the list of the edges and faces of the elements of the initial mesh and then to number these edges and faces.

It is thus easy to find the node numbers. Let N be the number of nodes to be defined on each edge and let NUM be the first free number (i.e. the number of vertices + 1 if they are nodes, or 1 otherwise) and let j be the global number of the edge under consideration; then the following number is assigned to the first node on this edge $V = NUM + (j - 1) \times N$ after which the others are numbered sequentially.

There still exists an ambiguity in the numbering, related to the direction in which the edge runs. To overcome this, it suffices to take the directional sense of the edge into account in the previous operation (for example, a solution is to number the node on the edge which is closest to its endpoint with the smallest number, first). This remark is only valid when $N > 1$.

To find the appropriate number for a node located on a face, the same

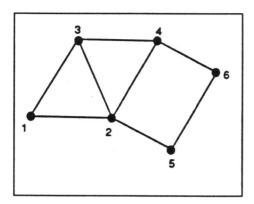 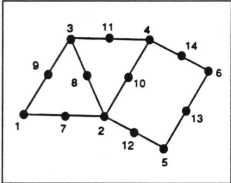

Figure 13.27: *Definition of the nodes.*

procedure is followed by considering the triangular faces and quadrilateral faces separately (as the number of nodes varies according to the type of the face investigated).

The same problem of ambiguity occurs when defining the list of the nodes associated with each element, when this number is greater than 1, due to the way the faces are described. To overcome this difficulty, it suffices to take the directional sense of the face when considering the element and the position of the vertex in this face with the smallest (or the greatest) number, into account in the previous operation.

Once the nodes of the edges and faces are known, the internal node numbers are assigned sequentially (there is no interaction between the internal nodes from one element to another).

Remark 13.10 : This operation of adding nodes does not know their exact position, but only their relative positions. The coordinates of the nodes are, in fact, determined at a later step (the definition of finite elements by themselves). □

Remark 13.11 : The addition of nodes is performed at the mesh conception level in order to find an "optimal" renumbering (cf. section 13.10). The addition of nodes is done once the mesh of the entire domain has been obtained, i.e., after all the operations concerning the creation or modification of the partial meshes used for the construction. □

13.10 Renumbering

In order to minimize the storage size required by the matrices computed during the numerical approximation, or the number of coefficients per row (column) of these matrices, or the number of operations required to solve the final system, it can be useful to renumber the nodes, or even elements, of a mesh.

This operation can also be motivated by the nature of the solution method [Lascaux,Theodor-1986]. For a frontal method, an adequate element numbering minimizes the CPU time by decreasing the width of the front. This is also true if the solution method involves transfers from secondary memory to main memory.

There are numerous methods available to perform this type of operation. We will describe the Gibbs method, which is completely automatic, in detail, and outline some other approaches.

In order to describe this algorithm, we introduce (or recall) the following notation:

- T_h denotes the triangulation,

- T_j is the element j of this mesh,

- NE is the number of elements of T_h,

- NOE is the number of nodes of T_h.

13.10.1 Renumbering of nodes

The renumbering of the nodes by the Gibbs method formally consists of following three steps [Gibbs et al. 1976]:

1. the search for a good point of departure (to initialize the algorithm),

2. the optimization of the numbering *descent*,

3. the numbering using the Cuthill MacKee algorithm, a numbering which can be reversed [Cuthill,McKee-1969].

These three phases are now described. Let us first define the additional notation below:

- $x, y, ...$ are the *nodes* of the triangulation T_h,

- y is a *neighbour* of x if there exists an element T of $\mathcal{T}_h$ such that x and y are nodes of it,

- the *degree* of a node x is its number of neighbours, it is denoted as $d(x)$,

- the neighbours of x constitute $N_1(x)$, *level 1* of its *descent*,

- the neighbours of the nodes of $N_1(x)$, not yet referenced, form $N_2(x)$, level 2 of the descent of x (a node appears at most once in $N_k(x)$ and thus cannot be present in any other level),

- the collection of levels $N_k(x)$ forms the *descent*, denoted by $D(x)$ of node x,

- *depth p* of the descent corresponds to the number of levels,

- *graph G* of the nodes indicates the connections between the nodes (in terms of neighbours); figure 13.28 shows the graphs corresponding to a triangle and a quadrilateral,

- graph G contains one or more connected components denoted by C_k.

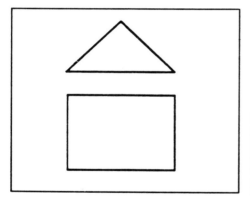 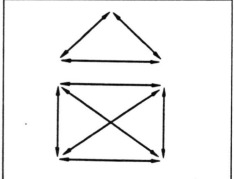

Figure 13.28: *Graphs associated with a triangle and a quadrilateral.*

The three phases consist therefore of:

1. **The search of an initial guess** :

    Node x is selected from the nodes located on the contour of the domain, in such a way that $d(x)$ is minimal.

13.10. RENUMBERING

Let $D(x)$ be its descent and $N_p(x)$ its last level; then the algorithm used is as follows:

- Begin
- For all nodes y of $N_p(x)$, in an increasing order of degree $d(y)$, compute $D(y)$ and its depth $p(y)$,
- If $p(y) > p(x)$ then select y, set $x = y$ and go to Begin,
- Else test a new y.

This results in a node y, the extremity of a *pseudo-diameter* whose other extremity, x, is a point in the last level of the descent of y such that its degree is minimal.

Phase two of the method is then started:

2. **The optimization of the numbering descent** : Let:

x and $D(x) = \{N_1(x), ..., N_p(x)\}$ and

y and $D(y) = \{N_1(y), ..., N_p(y)\}$

be the descents of the points of the pseudo-diameter, then we would like to construct $D = \{N_1, ..., N_p\}$ which will provide the new nodal numbering. This process is split into several phases:

- Loop for $w = 1$ to NOE
  Construction of the couples (i, j) associated with w such that:
  (i, j) is formed if $w \in N_i(x)$ and $w \in N_{p+1-j}(y)$
  End of loop.
- N_i is formed from these couples (i, j) in the following manner:
  Loop for $w = 1$ to NOE
  If (i, j) corresponding to w exists:
  place w in N_i and remove w from graph G of the nodes.
  End of loop.

If $G = 0$ go to the final step (Numbering of C.McK.)

Else the t connected components C_k of graph G are classed according to their cardinal numbers and we try to distribute the remaining nodes in levels N_i as follows:

- Loop for $k = 1$ to t

- Loop for $m = 1$ to p, the levels of descent
  Calculate:
  $n_m = \text{card}\{N_m\}$
  $h_m = n_m + \text{card}\{nodes \ \ of \ \ C_k \in N_m(x)\}$
  $l_m = n_m + \text{card}\{nodes \ \ of \ \ C_k \in N_m(y)\}$
  End of loop.

  Let $h_0 = \max_{m=1...p}\{h_m \ \ such \ \ that \ \ h_m - n_m > 0\}$ and
  $l_0 = \max_{m=1...p}\{l_m \ \ such \ \ that \ \ l_m - n_m > 0\}$

  - if $h_0 < l_0$ then: place each node of component C_k in level N_i defined by number i of couple (i,j)
  - else if $l_0 < h_0$ and if $\max_{i=1..p}\{cardN_i(x)\} > \max_i\{cardN_i(y)\}$ order i, otherwise order j.

  End of loop.
  In this way a well balanced distribution of the nodes is obtained in the different levels.

  - The Cuthill MacKee algorithm can now be implemented using these results.

3. **The numbering by the Cuthill MacKee algorithm**, may be reversed:

   If $d(y) < d(x)$, interchange x and y and reverse the associated levels:

   $N_{p-i+1} = N_i$ for $i = 1...p$

   Let us denote $NEW(x)$ to be the new number corresponding to the old value, n, of node x. Set:

   $NEW(x) = 1$

   Initialize: $n = 1$ and $N_0 = \{x\}$

   We then loop over the p levels of the descent:

   - Loop for $k = 0$ to $p - 1$
     a) While there is a w, the smallest number of the nodes of N_k preceding the neighbours not yet renumbered,
     do: $C(w) = neighbours(w) \bigcap N_{k+1}$
     b) while there exists s of the smallest degree of $C(w)$, not yet renumbered,
     do: $n \Leftarrow n + 1$ and $NEW(s) \Leftarrow n$

     - If a) interchanges x and y and b) chooses $N_i(x)$ for C_1

13.10. RENUMBERING

– or if a) does not interchange x and y and b) chooses $N_i(x)$ for C_1

End of loop.

Thus :

- Loop for $i = 1$ to NOE
  $s \Leftarrow NEW(i)$
  $NEW(i) \Leftarrow NOE + 1 - s$
  End of loop

this is the reverse of the renumbering, of interest due to the following result: the profile width is globally better, or at least as good, as the width before the reverse.

This algorithm is very effective for reducing the profile of a matrix, i.e., to minimize the difference between the node numbers of an element.

To illustrate this property, some examples are now displayed. Firstly, let us introduce the following definitions:

The *total profile* is the sum of the number of columns (resp. rows) included between the first non-empty column (resp. row) and the diagonal (in the case of a symmetric matrix).

The *average profile* is the quotient of this value and the number of columns (rows).

The *width* (or bandwidth) is the maximum difference between the first non-empty column (row) and the diagonal (in the symmetric case as before). This width is the quantity which needs to be minimized in the case of a direct solution method.

Mesh	1	2	3	4	5	6
Number of nodes	229	908	1559	61	3696	2056
Initial total profile	19592	42564	604714	1024	2266436	373601
Initial average profile	85.55	46.87	387.8	16.7	613.21	181.713
Initial bandwidth	217	897	1536	57	1183	2036
Final total profile	2637	25844	64931	379	436632	284920
Final average profile	11.51	28.46	41.64	6.21	118.13	138.580
Final bandwidth	18	45	66	13	233	225

Table 13.1 : *Profile before and after renumbering.*

Mesh	6	7
Number of nodes	2056	2056
Initial total profile	373601	284920
Initial average profile	181.713	138.580
Initial bandwidth	2036	225
Final total profile	284920	284883
Final average profile	138.580	138.562
Final bandwidth	225	225

Table 13.2 : *Renumbering process applied twice.*

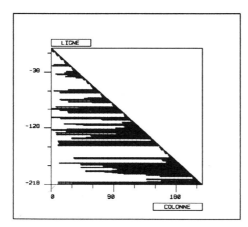
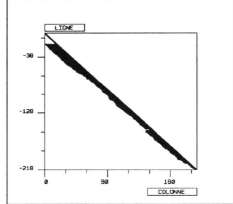

Figure 13.29: *Matrix occupation associated with the mesh.*

To summarize the efficiency of this method, its application to different meshes with various complexity has been shown. Example 1 is that of the mesh in figure 13.7 (right-hand side); case 2 corresponds to figure 13.19, the results are shown in figures 5.10 (left) (3), 5.10 (right) (4), 5.11 (left) (5), 5.11 (right) (6); case (7) deals again with the result of case (6) and runs the algorithm on it, the result remaining unchanged. Figure 13.29 shows the non-zero coefficients of the matrix associated with a mesh (that of above example 1) before and after renumbering.

Remark 13.12 : No data is required. □

Remark 13.13 : Contrary to this automaticity, a "special" numbering is not easy to obtain; to achieve this, one has to select a different method or to govern the algorithm oneself. □

13.10.2 Renumbering of elements

The renumbering of the elements is connected to the node numbering, thus, it is advisable to do, or simulate, it before this operation.

The following algorithm is proposed:

$num = 0$
$new(j) = 0$, for $j = 1$ to NE

- Loop for $i = 1$ to NE

    - for $T_j \in \mathcal{T}_h$ do:
      if T_j contains node i then:
      if $new(j) = 0$ then $num = num + 1$ and $new(j) = num$
      else $new(j)$ is already known
      End of loop.

End of loop.

Other methods exist, some of which are described below.

13.10.3 Other methods

To ensure some particular properties concerning the node or element numbering of a mesh, it can be useful to know other renumbering methods. While they are, in general, less efficient than the Gibbs method, they can be of use in certain situations.

- **Grooms method**: Of the iterative type, this method [Grooms-1972] consists of starting from the initial numbering and considering the longest row of the matrix, decreasing the difference between the first column of this row and the diagonal. The scheme is as follows:

Let NOE be the number of nodes of a mesh, and let α and β be two values.

  - $NTEST = \frac{NOE}{\alpha}$
  - $l = 0-$

-A- Loop k for 1 to $NTEST$

  * find the longest row of the matrix (it corresponds to the element for which the difference between the node numbers is maximal), and let L_1 and L_2 be the associated column and row indices,

* $INC = \frac{(L_2-L_1)}{\beta} - l$
* if $INC = 0$, the algorithm has converged.
* else, loop for $m = 0$ to $INC - 1$ (the numbering is compressed):
  · node $L_1 + m + 1$ replaces node $L_1 + m$,
  · node $L_2 - m - 1$ replaces node $L_2 - m$,
* end of loop m.
* node L_1 replaces node $L_1 + INC$,
* node L_2 replaces node $L_2 - INC$,
- End of loop k.
- $l = l + 1$
- go to -**A**-

The value α controls the number of tests in terms of the number of nodes in the mesh (a value of about 20 is advised), the number β interacts on the result (a value of about 2 is advised).

While it is relatively time consuming, this algorithm can be used for improving a numbering which is already quite good.

- **Lipton-Tarjan method** This method [Lipton,Tarjan-1979], [Lipton et al. 1979] relies on the notion of a graph and is based on a graph separation technique. For example, we find a discussion about this method in [Roman-1980], which consists of separating the problem into sub-problems of the same type but of smaller size.

- **Akhras and Dhatt method:**

Let i be a node and V_i the set of its neighbours, we consider the numbers of the nodes in V_i, and evaluate the following quantities:

- S the sum of the numbers of the nodes in V_i,
- $P = \frac{S}{card(V_i)}$ the average of the numbers of the nodes in V_i,
- S_p the sum of the smallest and greatest number of the nodes in V_i.

The algorithm is based on the analysis of the optimal numbering of the regular mesh of a quadrilateral and the values of P and S_p associated with this mesh. P and S_p are increasing; thus, one tries to construct a numbering satisfying these properties in the domain under consideration. Being iterative, this algorithm comprises two phases:

13.10. RENUMBERING

- the nodes are renumbered (using one or several iterations) in such a way that P and S_p are increased;
- the result is renumbered in such a way that P and S are increased (phase 1 can then be restarted).

This method [Akhras,Dhatt-1976] is quite performing and automatic; in contrast, the time required may be considerable.

- **Frontal method**: This method consists, in essence, of radiating from a starting front.

Let F_o be a set reduced to one node or a collection of nodes (generally chosen on the contour of the domain), then the algorithm can formally be written as follows:

- $k = 0$

-**A**- for $i = 1, card(F_k)$ (the number of nodes of F_k):
  - Loop i over the elements containing the ith node of F_k:
    * construction of V_i, the set of numbers of the nodes neighbouring node i, not yet renumbered,
    * Loop j over the nodes of V_i
      · computation of $NBVOIS(j)$, the number of neighbours of node j,
    * End of loop j.
    * renumbering of the nodes of V_i according to $NBVOIS(j)$, the number of neighbours (in the increasing or decreasing order of $NBVOIS$)
  - End of loop i.

The set V_i forms the new renumbering front, thus:

- $k = k + 1$
- $F_k = V_i$
- if $card(F_k) > 0$ go to -**A**-,
- else, the process is completed (the resulting numbering may be reversed)

This method of node renumbering depends strongly on the choice of the initial front and therefore requires the user's directive. In addition, numerous variants exist.

- **Element colourization method**: This method consists of, using ideas similar to the four colours theorem, renumbering the elements of a mesh by creating packets of elements such that any neighbour of an element in a given packet is not located in the same packet. In the case where only two packets are created, this method clearly looks like the "Red-Black" method [Melhem-1987] which permits separation of nodes into two disjoint sets. Anyway, the algorithm is based on the search for the neighbours of an element with the aim of separating these elements. The main application of this method is the numerical solution of a problem on a vector computer, but this idea can be applied to different areas.

- **Nested dissection method**: The problem is recursively split into two sub-problems. Initially developed for structured meshes (grids) [A. George-1973], this method extends [George,Liu-1978] to arbitrary meshes.

As a conclusion to the renumbering methods, we note that each method has advantages and weaknesses. In practice, numerous authors propose combining several methods. A detailed discussion can be found in [Marro-1980] concerning some of these renumbering methods and, in particular, the combination of different methods in such a way that their negative features are avoided and their respective qualities are retained.

13.11 Controls and graphics

The mesh analysis is a crucial and difficult point when investigating meshes with a large number of elements, especially in the three-dimensional case. Two complementary approaches can be followed: the graphic visualization and the numerical check.

For efficiency, the graphic visualization must offer numerous tools to allow the easy examination of this or that part or characteristic of the mesh under consideration. In addition, these tools must be incorporated in a system which must be as user-friendly as possible.

These programs are often large (in terms of lines to be written) and their programming can be very complex.

The numerical verification of the meshes is based on the computation of quantities associated with them. For example, it can be of interest to be sure that the surfaces or volumes of elements are all positive. Thus, the availability of histograms showing the distribution of elements as a function of their quality ($Q = \frac{\rho}{h}$ or a definition involving some criteria of physical

13.11. CONTROLS AND GRAPHICS

nature connected to the problem to solve) gives a quick understanding of the mesh under consideration.

The purpose of this section is to make a tentative list of all the tools required when considering the graphic approach. Some information will be given concerning the algorithms needed and their programming. Lastly, the graphic processors included in the Modulef library[George et al. 1987] will be used to illustrate a real application of this type of graphic background.

13.11.1 Definition of a graphic environment

Schematically, all drawings result from several manipulations at different levels. In the present case, it corresponds, starting from one or several mesh data structures to visualize, to:

- Reading the appropriate structure to access the information it contains;

- Exploiting these values to extract the information which is strictly useful with regard to the plot;

- Translating this information in accordance with the graphic software at our disposal;

- Requesting the plot by stipulating the choices related to the desired visualization mode.

The scheme in figure 13.30 outlines this process. The latter includes several layers concerning different kinds of treatments. The general problem is to conceive and write these layers in such a way that a set of visualization programs is obtained which is general-purpose, evolutive and easy-to-use. These imperatives are linked, on one hand, to the variety of the drawings desired and, on the other hand, to the necessity to foresee the arrival of new inventions and, lastly, to obtain a quick and efficient analysis of meshes and, in addition, provide the user with an easy way of specifying his choices.

- **Exploitation of the data structure** : Layer [1] is directly dependent on the chosen mesh data structure, exploits this structure and gives access to the descriptive values of the mesh.

- **Preliminary computations** : Layer [2] considers the values resulting from the previous layer and deduces the necessary values (computation of the extrema of the object, definition of the visualization conditions, etc.) to constitute the set of values defining the plot (list

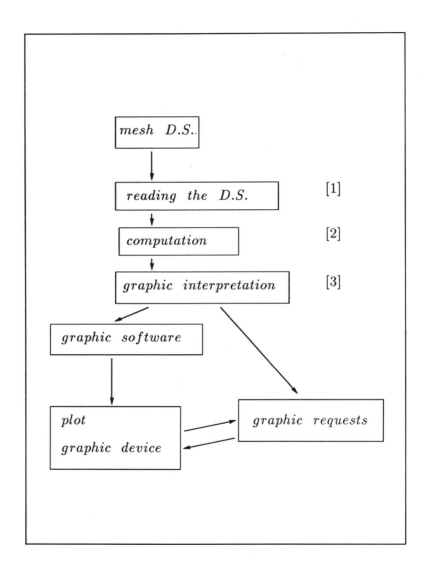

Figure 13.30: *Formal plotting scheme.*

13.11. CONTROLS AND GRAPHICS

of the elements to be displayed, list of the faces, edges, set of the attributes to associate with the picture (characters, numbers, legends, colours, etc.)).

- **Graphic interpretation** : Layer [3] interprets the plot request with regard to the graphic software available[3].

- **(Interactive) plot modifications** : Layer [4] allows the user to request a new drawing by manipulating the previously obtained output with the help of an input facility (reticule, digital table, mouse, etc.).

13.11.2 Graphics for two-dimensional meshes

In two dimensions, the availability of the following features seems advisable:

- plot of the elements of the mesh;

- "shrink" of these elements; a shrink is a contraction of a given ratio of each element about its centre of gravity. This is a way to check, at a glance, the conformity of a mesh (according to definition 1.1);

- display of certain mesh items (its boundary, some particular elements, etc.);

- obtain a plot displaying some indications (point, node, element, reference, sub-domain, ..., numbers.);

- zoom-in on a given region of the mesh;

- manipulate different views of the same mesh in a general way (return to a view seen previously, etc.), or change from one mesh to another.

13.11.3 Graphics for three-dimensional meshes

In three dimensions, the previous features are recommended and, because of the increased difficulty, the following additional features are pertinent:

- display of the single surfaces of the mesh;

- "shrink" of these;

---

[3]Unfortunately, several graphic norms exist, among which are GKS and PHIGS. The choice of the one or the other of these norms is not easy nowadays. One may hope that this point will be cleared up in the very near future.

258   CHAPTER 13. MESH TRANSFORMATIONS

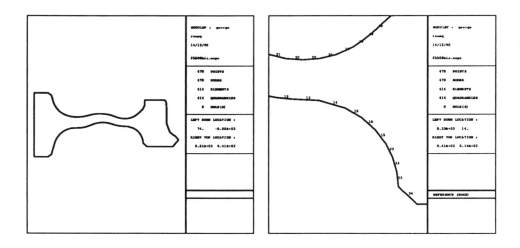

Figure 13.31: *Geometrical boundary of a mesh and a zoom of it with physical numbers.*

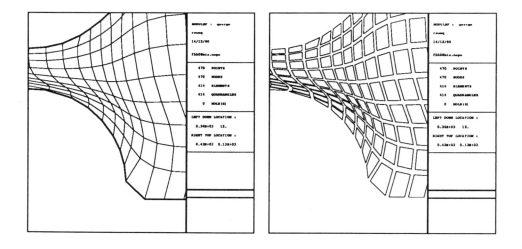

Figure 13.32: *Mesh and a mesh "shrinked" around a region.*

13.11. CONTROLS AND GRAPHICS

- obtain a plot displaying some indications (point, node, element, reference, sub-domain, ..., numbers.);

- zoom-in on this or that region of the mesh or its surface;

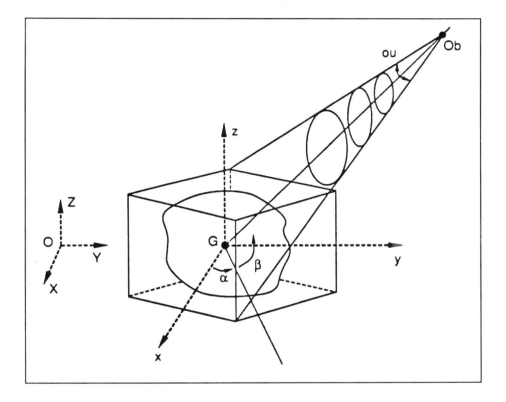

Figure 13.33: *Observation modes.*

- change the observation conditions (turn around the object, move away, move closer to it, perform some cut, ...). The observation is defined via:
    - cartesian coordinates by:
        * the position of the observer (Ob),
        * the position of the view point (G).
    - spherical coordinates by (figure 13.33):
        * a sight angle (α),
        * an elevation angle (β),

* an aperture angle (Ou).

The change from one system to another must be possible automatically. Thus, an automatic observation mode must be available to avoid the definition of the relevant values by the user: the domain is enclosed in a box, its extrema are used to compute a *reasonable* observation position. Next, a different view is possible by modifying some parameters.

- use of colours for better readability (see the colour insert).

When displaying the surface of an object, it is indispensable to take the notion of visibility into account (i.e., to only display the visible (or invisible) parts). This visibility is linked to the position of the observer. The computations for knowing if an item is visible or not are not trivial: several algorithms may be involved. A simple method consists of computing, for each face, an oriented normal. If the latter is directed towards the observer, the face is called visible (this algorithm is very fast but only gives one approximate solution to the visibility problem). The methods based on the "algorithm of the painter" are more elaborate: the plot is displayed starting with the plot of the faces which are farthest from the observer, on a terminal with selective erasing, the final result includes only the visible parts of the object (this method generally provides good results).

To appreciate elements of a mesh, some of them can be selected and viewed by animation (the element chosen is submitted to movements so that it can be observed from different angles, analyzing its shape in this way).

Remark 13.14 : The use of graphic facilities is also useful in the evaluation, analysis and communication of computational results (velocity, pressure, temperature distributions, etc.). These tools are used for the efficient display of the associated values (scalars, vectors, tensors, etc.). □

Remark 13.15 : Some codes allow for graphic interactive mesh modification. This feature can, in certain situations, be used to improve the mesh locally. □

The Modulef library [George et al. 1987] proposes a suite of graphic preprocessors which are machine independent and are suitable for the visualization of meshes. Most of the examples included in this book have been produced using the modules of this suite.

13.11. CONTROLS AND GRAPHICS

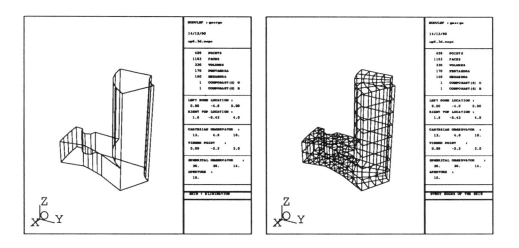

Figure 13.34: *Plot of wired lines of the skin with and without the removal of the coplanar edges.*

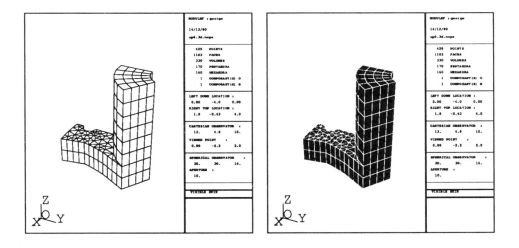

Figure 13.35: *Visible skin, facets colored in black and white.*

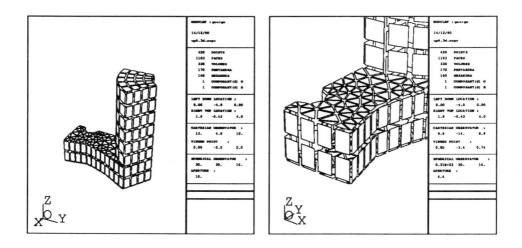

Figure 13.36: *Visible skin in a shrinked mode, zoom and rotation.*

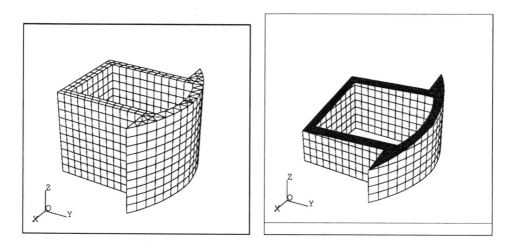

Figure 13.37: *Mesh of a magnetic head and cut by a plane.*

Chapter 14

Some mesh generation packages

14.1 Introduction

First, this chapter presents the different facilities in the Modulef library concerning the generation and modification of two-dimensional and three-dimensional meshes [George-1989a]. A brief description of this software gives a comprehensive and practical illustration of a possible way to implement the main ideas introduced throughout this book. Furthermore, it should be mentioned that the author assumes the responsibility of the mesh generation part of the Modulef library for which he wrote, or supervised the writing of, numerous programs.

Next, a system using the C.A.D.[1] technique is introduced, suitable, on one hand, for data capturing when creating a two-dimensional mesh and, on the other hand, for the creation and the interactive manipulation of such meshes. This package [Hecht,Saltel-1989] has also been developed at INRIA for the creation of meshes compatible with those resulting from the Modulef library.

Obviously, numerous other packages exist offering important functions in the field of mesh creation. Some of these will be mentioned briefly.

14.2 The Modulef library (mesh generation part)

The Modulef library proposes various algorithms for two-dimensional mesh creation: the methods effectively implemented correspond to a manual ap-

---

[1]C.A.D.: Computer Aided Design.

proach (cf. chapter 5), a transport-mapping function (cf. chapter 6), an advancing front technique (cf. chapter 10), a Voronoï method (cf. chapter 11), and finally, a multiblock method (cf. chapter 9). In the three-dimensional case, we find the manual and semi-automatic approach (cf. chapter 5), the transport-mapping-deformation technique and the multiblock approach (cf. chapters 6 and 9)). A large suite of programs suitable for modification (symmetry, refinement, etc.) is also available. Moreover, a set of preprocessors facilitates the practical use of the programs in most cases.

The main tasks when creating a mesh using the Modulef philosophy[2] is summarized below. For a more technical description, consult [George-1989c] and [George-1989a] from which this summary is taken.

- For the **two-dimensional case** it corresponds to:

    - creating meshes using point, line, contour, coarse mesh, ..., data and adequate algorithms;
    - modifying existing meshes and combining them to obtain, step by step, the final mesh with which a *NOPO* type data structure is associated; this structure is stored in a sequential access file.

To achieve this:

   - Primal sub-sets are located by eliminating all sub-sets whose meshes can be deduced by symmetry, translation, ..., or more generally by any geometrical modification;
   - In addition, primal sub-sets due to the physical constraints of the problem are recorded;
   - For each of these sub-sets, the most appropriate module is selected;
   - For each of these modules, the nature of the data it uses is examined;
   - The data is constructed from the notions of points and lines;
   - The desired mesh generator(s) is(are) activated;
   - The relevant symmetries, translations, rotations, ... of the primal sub-sets are applied;
   - Sub-meshes are pasted together;

---

[2]This description corresponds to the 1990 version of the software.

14.2. THE MODULEF LIBRARY (MESH GENERATION PART)

- The previous processes are repeated until a mesh is obtained covering the complete domain;
- Some local or global modifications (refinement, regularization, ...) can then be performed;
- The non-vertex nodes, if any, are defined and the vertices are removed from the node list if necessary;
- Elements and/or nodes can be renumbered;
- The result is then stored on file; the quality of the mesh can be checked by plotting it.

- For the **three-dimensional case** the operations are of the same nature and therefore correspond to:

  - creating meshes from the point, line, contour, two-dimensional mesh, coarse mesh, ..., data and adequate algorithms;
  - modifying previously created meshes and combining them to obtain, step by step, the final mesh with which a $NOPO$ type data structure is associated; this structure is stored on a file with sequential access.

To achieve this:

- Primal sub-sets are located by eliminating all sub-sets whose meshes can be deduced by symmetry, translation ..., or more generally by any geometrical or topological modification;
- In addition, primal sub-sets due to the physical constraints of the problem are recorded;
- For each of these sub-sets, the most appropriate module is selected;
- For each of these modules, the nature of the data it uses is examined;
- The data is constructed from the notions of points, lines, faces, ...;
- The desired mesh generator(s) is(are) activated;
- The relevant symmetries, translations, rotations, ... of the primal sub-sets are applied;
- Sub-meshes are pasted together;

- The previous processes are repeated until a mesh is obtained covering the complete domain;
- Some local or global modifications (refinement, regularization, ...) can then be performed;
- The non-vertex nodes, if any, are defined, and the vertices are removed from the node list if necessary;
- Elements and/or nodes can be renumbered;
- The result is then stored on file; the quality of the mesh can be checked by plotting it.

Having done this analysis, one of the following approaches can be used from a practical point of view:

- use of general purpose preprocessors;
- use of specific purpose preprocessors;
- classical call of modules.

The first two ways correspond to a conversational approach, the last corresponds to "batch" processing.

14.2.1 Conversational use

Background

The Modulef library [Bernadou et al. 1988] is organized into several sets of programs or (sub-)libraries. Each library contains programs corresponding to a specific type of application. In this respect, the general purpose utilities are located in library $UTIL$, those devoted to the manipulation of the data structures are in $UTSD$, the mesh generators and the modules dealing with two-dimensional meshes are found in $NOP2$, those dealing with three-dimensional meshes are in $NOP3$, the programs not devoted to any specific application are in $NOPO$, ..., the main programs are located in $PPAL$, the utilities used by the conversational system are in $CONV$, etc.

For a two-dimensional analysis, it is necessary to link libraries $CONV$, $NOP2$, $NOPO$, $UTSD$, $UTIL$ and $UTIF$ (general machine-dependent utilities).

In three dimensions, the same type of link is required, but in this case library $NOP3$ is used instead of $NOP2$.

The desired main program (contained in $PPAL$) can then be executed.

14.2. THE MODULEF LIBRARY (MESH GENERATION PART)

General purpose preprocessors

In the two-dimensional case, processor APNOXX activates all the mesh related modules (creation and modification), and in the three-dimensional case, processor APN3XX gives access to all the modification modules and some creation modules suitable for simple cases.

These two programs rely on *KEY-WORDS* and require the input of a data file containing the requests specified in this form (key-word and associated values). More specifically, there is an option available to create the data file (or commands file), and another option to read this file (previously created) and to execute the requests contained in it.

As an illustration, the possibilities present in processor APNOXX will now be described. This main program calls the general module APNOPO[3] which then selects the one or other basic module. A key-word is associated with each possibility, a complete list of which is given below:

1. Preliminary preparation

    - COUR : definition of functions describing curves, if any,
    - POIN : definition of characteristic points,
    - LIGN : definition of characteristic lines.

2. Mesh generation modules

    - QUAC : call mesh generator *QUACOO* which corresponds to a transport-mapping method for the case of a generalized quadrilateral,
    - TRIC : calls mesh generator *TRICOO* which corresponds to a transport-mapping method for the case of a generalized triangle,
    - TRIA : calls mesh generator *TRIGEO* which corresponds to an advancing front method,
    - TRIH : calls mesh generator *TRIHER* which corresponds to a Voronoï method,
    - MANU : calls mesh generator *CONOPO* which corresponds to a manual description,
    - OBJE : calls mesh generator *CONOPO* in the case of a single element,
    - BARR : calls mesh generator *CONOPO* to create segments.

3. Modification modules

    - TRAN : calls module *MODNOP* to perform a translation,
    - ROTA : calls module *MODNOP* to perform a rotation,

---

[3]The first version of this general module, module APMEFI, is due to [Perronnet-1974]; it was implemented in 1974.

- SYMD : calls module *MODNOP* to perform a symmetry w.r.t. a line,
- DILA : calls module *MODNOP* to obtain a dilation,
- Q4T : calls module *QUATRI* to cut each quadrilateral into four triangles,
- RETR : calls module *RETRIN* to split each element into sub-elements of the same type,
- AFFL : calls module *AFFNOP* to refine a mesh locally,
- REGU : calls module *REGMA2* to regularize a mesh,
- AIGU : calls module *AIGUNO* to create a non-obtuse mesh
- NUME : calls module *MODNOP* to modify attributes (reference or sub-domain numbers),
- RECO : calls module *RECOLC* to paste two meshes together,
- ADPO : calls module *ADPNOP* to define the nodes,
- RENE : calls module *GIBBS* to renumber the elements of a mesh,
- RENC : calls module *GIBBS* to renumber the elements and the nodes of a mesh,
- COUL : calls module *COULEU* to renumber the elements using different colours.

4. Save and End

- SAUV : calls module *SAUVER* to store any mesh residing in core on file,
- FIN : request to terminate the job.

5. Utilities

- LIRE : request to read any other data file (created using the same format),
- CONT : request to suppress the data coherence checks,
- INTR : calls module *SDREST* to load a mesh stored on file in core,
- TUER : calls module *TUERSD* to remove one or more data structures from core,
- IMPR : calls module *IMNOPO* to print the contents of a data structure,
- DESS : calls module *TRNOPO* to plot a mesh,
- INFO : calls module *INFONO* to interrogate a mesh data structure,
- MENU : request to obtain a list of key-words available,
- ? or whatever : corresponds to key-word MENU.

In the case of an error, key-word MENU is selected.

Consult [George-1989a] for a description of processor APN3XX which calls module APNOP3 (for the three-dimensional case).

14.2. THE MODULEF LIBRARY (MESH GENERATION PART)

Specific purpose preprocessors

Having the same background, specific processors call one of the modules of the library. They fall into two groups:

- For the modules requiring very little data, the processor starts by requesting the necessary values and then activates the module.
- For the opposite case, we return to the presentation of processor APNOXX; a first phase consists of creating a data file which is read in the second phase after which the requests contained in it are executed.

For this configuration, the different processors are:

- CHARXX : (CHARPO CHARPL) To define a junction between two beams or two shells.
- COLIXX : (COLIBR) To create a $3D$ mesh from blocks.
- MA23XX : (MA2D3D) To construct a $3D$ mesh by stacking a $2D$ mesh.
- TN2DXX : (TN2D3D) To create a surface in R^3 from a $2D$ mesh.

for the mesh generators and,

- ADPNXX : (ADPNOP) Addition of non-vertex nodes.
- AFFNXX : (AFFNOP) Local refinement in 2 dimensions.
- AIGUXX : (AIGUNO) Local modification of a mesh to obtain a non-obtuse mesh (2D only).
- COMPXX : (COMPRE) Compression of a D.S. NOPO so that the nodes and the points coincide.
- COULXX : (COULEU) Renumbering of elements using different colours.
- DETRXX : (DTRI3D) Splitting the elements of a $3D$ mesh into tetrahedra.
- ELIMXX : (ELIMNO) Removal of one or several elements from a mesh.
- GEODXX : (GEODET) Mesh check by computing the element surfaces and volumes.
- GIBBXX : (GIBBS) Renumbering of the nodes and elements of a mesh.
- MOD1XX : (MODNO1) Manual modification of values contained in a D.S. NOPO.
- MODNXX : (MODNOP) Modifying a mesh by geometric transformations.
- RECOXX : (RECOLC) Pasting two meshes together.
- RETRXX : (RETRIN) Splitting elements of a mesh into sub-elements of the same type.
- REFEXX : Graphic - (REFEGE) Display of $3D$ meshes, cuts, checks, modification of numbers and geometrical modification.
- TRNOXX : Graphic - (TRNOPO TRGEOM) Display of any mesh.

for the modification and graphic purpose modules.

14.2.2 Classical use

In the event that the conversational approach is either too long or simply not possible, the modules in the library can be called directly.

Background

The link required is the same as above except that library $CONV$ is no longer used.

A calling program is to be written in Fortran 77 as follows:

- COMMON M(LM)
  Declaration of the super array M of length LM words
- Declaration of "functions", "externals", tables used, ...
- CALL INITI(M,LM,IMPRE,NNN)
  Initialization of all MODULEF main programs
- .... the input parameters of the module ....
- CALL MODULE(...)
  calling the module
- STOP
- END

If the module (MODULE above) uses functions or subroutines, these must be written, compiled and linked.

General modules

For the two-dimensional case, and using the above scheme, module APNOPO is called in the following manner:

- CALL APNOPO(M,M,FFRONT)

The only arguments of this module are the super array M and the formal name of an optional mapping function. The data is contained in a data file as specified above (key-words and associated values). The required specifications to create such a file are found in [George-1989a].

For the three-dimensional case, and using the same scheme, module APNOP3 is called as follows:

- CALL APNOP3(M,M,XYZ23,XYZ33)

The only arguments of this module are the super array M and the formal name of two optional subprograms. As seen in the $2D$ case, the data is contained in a data file using the same format (see also [George-1989a]).

14.2. THE MODULEF LIBRARY (MESH GENERATION PART)

Specific modules

In order to call one of the basic modules of the library, the above scheme must be followed with the help of reference [George-1989a]. Furthermore, as the programs of the Modulef library are available in source code, the reader can refer to it directly.

14.2.3 Open feature of the package

As indicated above, the source programs of the Modulef library are accessible. Great care has been taken in the presentation of the modules to ensure facility and completeness. In particular, it is easy to know the list and meaning of the input and output parameters of each program.

In addition, a *procedure base* can be consulted interactively, providing information about the program hierarchy (for example, for a given calling program, one can access information concerning all the routines it calls: called program ⇔ calling program).

These original properties ensure this characteristic feature of the library called *an open library*. They have been conceived in such a way as to incorporate unforeseen cases. Two examples exhibiting this type of approach are given in chapter 15.

To illustrate the manner in which the modules are coded, we consider module RECOLC (pasting of two meshes) by listing the first lines[4].

```
      SUBROUTINE RECOLC(M1,M2,EPS,NFNOP1,NINOP1,NFNOP2,NINOP2,IOPT,IEB,
     +                 NFB1,NIB1,NFB2,NIB2,NFNOPS,NINOPS,NTNOPS,
     +                 NFBS,NIBS,NTBS)
C ++++++++++++++++++++++++++++++++++++++++++++++++++++++++++++++++++
C BUT : RECOLLEMENT DE DEUX MAILLAGES 'NOPO' ET EVENTUELLEMENT
C ---     DE DEUX STRUCTURES    ' B ' S' APPUYANT SUR CES MAILLAGES
C ++++++++++++++++++++++++++++++++++++++++++++++++++++++++++++++++++
C IN :
C --
C M1,M2       : LE SUPER TABLEAU ( 2 SONT POSSIBLES )
C EPS         : PRECISION RELATIVE POUR LA COMPARAISON DES SOMMETS
C NF(NI)NOP1(2) : FICHIER ET NIVEAU DES S.D.E. NOPO
C IOPT        : OPTION DE RECOLLEMENT DES FRONTIERES
C       0     : SANS VERIFICATION DES REFERENCES :
C               2 SOMMETS DE REFERENCES DIFFERENTES ==> RECOLLES ET REF 0
C       1     : VERIFICATION DES REFERENCES :
C               2 SOMMETS DE REFERENCES DIFFERENTES ==> NON RECOLLES
C IEB         : 0 : 2 NOPO SEULEMENT ,  1 :  2 NOPO ET 2 B
```

---

[4] While programs are written in french, print-out can be obtained in english.

```
C     NF(NI)B1(2) : FICHIER ET NIVEAU DES S.D.E. B (APPUYEES SUR LES NOPO)
C     NF(NI)NOPS(BS) : FICHIER ET NIVEAU DES S.D.S. NOPO ET B EVENTUELLE
C     NT(NOP,B)S : TABLEAUX ASSOCIES A LIRE SUR CARTES
C     ++++++++++++++++++++++++++++++++++++++++++++++++++++++++++++++++
C     PROGRAMMEURS : P.LAUG P.L.GEORGE F.HECHT F.PISTRE   : RELEASE 1987
C     ................................................................
```

14.3 EMC^2, a two-dimensional mesh editor

14.3.1 Introduction

The software package EMC^2 can be used for the construction of $2D$ meshes using a mesh and $2D$ contour editor. Basically, this software consists of three parts: one is dedicated to **data construction** (points, lines, arcs, contours, ...), another is devoted to **mesh generation** (by means of Voronoï type method) and the last is used for easy **mesh manipulation** (quadrangulation, symmetries, etc.). This software package is compatible with the Modulef code described above.

The aim of this section is to present the different capabilities of the EMC^2 software package briefly. For a more complete description, [Hecht,Saltel-1989] should be consulted.

14.3.2 General overview

The software is divided into three applications:

- the **CONSTRUCTION** application for the editing and creation of contours,

- the **PREP_MESH** application for editing and defining the mesh from contours and sub-domains,

- the **EDIT_MESH** application for the creation and editing of meshes (using interactive processes).

In each of these applications, several menus are available:

- a **GENERALITES** menu for specifying or changing the application, saving or restoring a data base, making soft-copies, re-running a session, quitting, etc.

- a **GESTION DE L'ECRAN GRAPHIQUE** menu,

14.3. EMC², A TWO-DIMENSIONAL MESH EDITOR

- a **CALCULETTE** menu to capture a numerical value or to obtain a value by interrogation,

- a **DESIGNATION** menu to identify or define the items to be used.

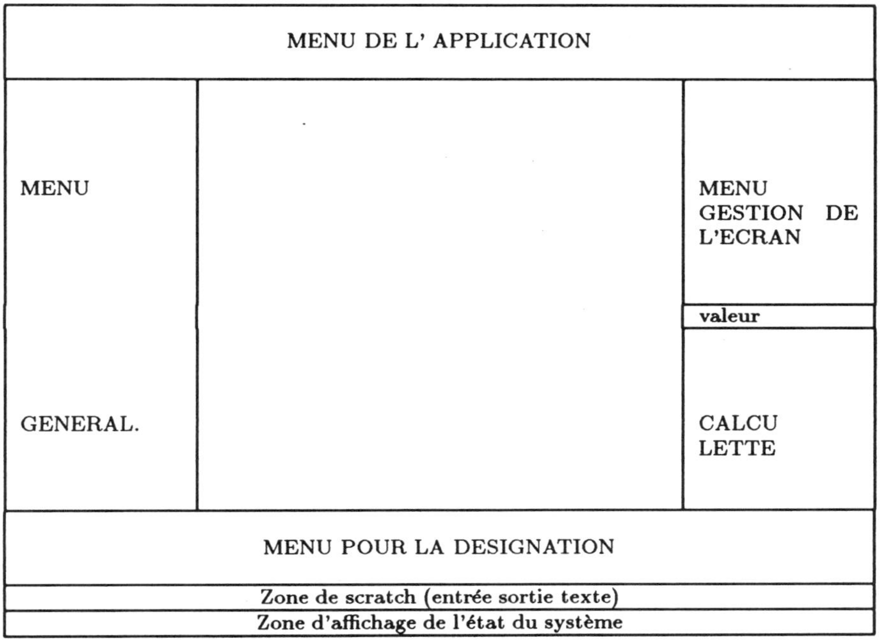

MENU DE L' APPLICATION		
MENU GENERAL.		MENU GESTION DE L'ECRAN
		valeur
		CALCU LETTE
MENU POUR LA DESIGNATION		
Zone de scratch (entrée sortie texte)		
Zone d'affichage de l'état du système		

Scheme 14.1 : *Screen presentation.*

Some menus are available for all the applications, and some are dependent on these. In practice, EMC² is an interactive software package which has a screen presentation with eight regions (see scheme): four menu regions, three print-out regions and the main region at the centre of the screen.

The CONSTRUCTION application

This application is used to define the geometrical contours of the domain in terms of points, segments, circle arcs and splines. Two additional entities have been introduced which are of interest in some cases: lines and circles.

All the entities, except splines, can be constructed with the help of elementary geometry theorems. When there are multiple solutions, the ambiguities are overcome by using the following heuristic: the designation points close to the curve tangency points are retained.

A spline is defined as a curve C^1 passing through the points of a list. It is closed if the first and last points are identical.

It is possible to duplicate all these entities by using the following affine transformations: symmetry, rotation, homothety, translation.

In addition, it is possible to:

- smooth the angles;

- cut the segments, arcs, splines by points, segments, arcs, splines;

- remove entities.

The application PREP_MESH

At this level, the connected components (denoted simply components) of the boundaries of the sub-domains of Ω are known.

This application defines the discretization of the entities defining the boundary of the domain and sub-domains (material boundaries), and also the reference numbers of the sub-domains, lines, points, in order to assign different physical data (for example: several boundary conditions or materials).

Moreover, it can be used to create a data file serving as an interface for the APNOXX mesh generator of the Modulef library, by defining the sub-domains by the list of components of their contour, where the first is the exterior component (the other components being components which are holes); internal lines or points can be specified in addition.

The application EDIT_MESH

The value of certain mesh generator parameters (specified by default) can be modified. The program creates a mesh of triangles of the sub-domains defined in the previous step, or of all the sub-domains if none were defined explicitly.

At this level, the mesh can be edited to:

- add internal vertices to the sub-domains;

- remove vertices, sub-domains;

- swap the internal edges of a quadrilateral formed by two triangles;

- move vertices;

- smooth a mesh or modify it in such a way that it is of type *Delaunay*;

14.3. EMC², A TWO-DIMENSIONAL MESH EDITOR

- modify a mesh to contain (only) quadrilaterals;
- transform the sub-domains by symmetries, rotations, homotheties, translations;
- modify the vertex, edge, sub-domain (region) physical attributes;
- save or restore a mesh (in different forms: *MESH* i.e. the internal structure of EMC², *NOPO* i.e. the Modulef structure, etc.);
- create a crack in the lines of the mesh (the points on the two parts of the line are duplicated);
- renumber the nodes to minimize the size of the matrix skyline, for finite element types P^1 or Q^1.

Rather than give a longer description of the EMC² software, we are going to give a complete example of its use (this example has been taken from [Hecht,Saltel-1989]).

14.3.3 A complete example

In this section we show how to obtain a mesh of a unit square containing a hole. We start by executing program EMC² and specifying the graphic device. We are then, by default, in the CONSTRUCTION application.

Creation of a unit square with a hole

1. press *POINT*[5] (designation menu);
2. press *PT_XY* (designation menu);
3. type on the keyboard: 0=0=1=0=0=1=1=1=.5=.5= (= can be replaced by return). *five points appear at the centre of the screen* (the scaling is assigned to 1);
4. press *VOIR_TOUT* to see the five points. The fifth will be used for the definition of the centrepoint of the hole;
5. press *SEGMENT* (construction menu at the top, that at the bottom is the item *SEGMENT* of the designation menu);
6. press *POINT* (designation menu);
7. press the endpoints of the four segments to be created in the graphic window: after each couple of presses a segment is created. This is why it is necessary to press eight times close to the eight points (0,0), (1,0), (1,0), (1,1), (1,1), (0,1), (0,1), (0,0), in this order;
8. press *CERCLE* (menu construction) then *CENTRE* (same menu);

---
[5] In practice, one has to press on the box *POINT* of the menu

9. press close to point (.5,.5) in the graphic window (note that we are still in the point designation mode);
10. press *RAYON* (designation menu),
11. type the keyboard 1/4= (or 1/4 return), a circle with centrepoint (.5,.5) and radius 1/4 appears. *Warning: this circle, as for the line, are only a support for the construction so we have to transform them explicitly into the arcs desired;*
12. press *ARC* of the designation menu to generate arcs;
13. press *IDEM*;
14. press *CERCLE* (designation menu) to identify the circle;
15. press the graphic window (as there is only one circle, only this one can be specified);
16. press *DETRUIRE* (general menu), as the circle is no longer useful, it is removed;
17. press the graphic window (the circle and arc disappear because they overlap). *warning: we are still in destruction mode;*
18. press *RAFRAICHIR* to refresh *(the arc is still visible).*

We have now completed the construction of the geometry (figure 14.1), and can continue to the PREP_MESH application.

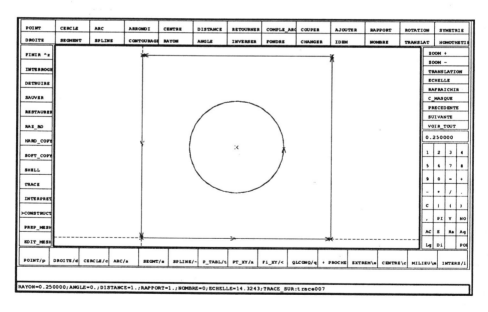

Figure 14.1: *The geometry of a square domain with a hole.*

Definition of the discrete form of the boundary

1. press *PREP_MESH* (general menu);
2. press *NB_INTERVAL* to define this value;

14.3. EMC², A TWO-DIMENSIONAL MESH EDITOR

3. type 4=;
4. press *TOUT* of the application menu (top) (all the elements are split into four intervals);
5. type 12=;
6. then press the graphic window close to the arc which is then split into 12 intervals (it it is not necessary to change from designation mode);
7. press on *SAUVER* of the menu general;
8. type the name of the file: "carre_troue" and return, in this way the file *carre_troue.emc2_bd* is created.

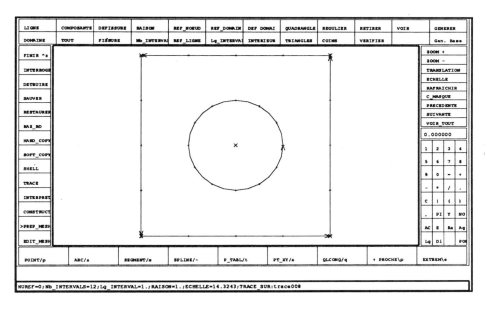

Figure 14.2: *The discretization of the lines of the domain.*

The file carre_troue.emc2_bd is listed below:

```
'-- TYPE I BD1 BD2 BD3 BD4 BD5 NBNO RAIS RFG RFD RF1G RF1D RF2G RF2D FIS'
'DROITE'     1 -1.000 0.0000 1.000 0.0000 0.0000 0.0000 0 0.0000 0 0 0 0 0 F
'DROITE'     2 -1.000 1.000 0.0000 0.0000 0.0000 0.0000 0 0.0000 0 0 0 0 0 F
'POINT'      3 0.0000 0.0000 0.0000 0.0000 0.0000 0.0000 0 1.000 0 0 0 0 0 F
'POINT'      4 0.0000 1.000 0.0000 0.0000 0.0000 0.0000 0 1.000 0 0 0 0 0 F
'POINT'      5 0.0000 0.0000 1.000 0.0000 0.0000 0.0000 0 1.000 0 0 0 0 0 F
'POINT'      6 0.0000 1.000 1.000 0.0000 0.0000 0.0000 0 1.000 0 0 0 0 0 F
'POINT'      7 0.0000 0.5000 0.5000 0.0000 0.0000 0.0000 0 1.000 0 0 0 0 0 F
'SEGMENT'    8 -3.000 0.0000 0.0000 1.000 0.0000 0.0000 5 1.000 0 0 0 0 0 F
'SEGMENT'    9 -3.000 1.000 0.0000 1.000 1.000 0.0000 5 1.000 0 0 0 0 0 F
'SEGMENT'   10 -3.000 1.000 1.000 0.0000 1.000 0.0000 5 1.000 0 0 0 0 0 F
'SEGMENT'   11 -3.000 0.0000 1.000 0.0000 0.0000 0.0000 5 1.000 0 0 0 0 0 F
'CERCLE'    12 0.2500 0.5000 0.5000 0.0000 0.0000 0.0000 2 1.000 0 0 0 0 0 F
'ARC'       13 -2.000 0.5000 0.5000 0.7500 0.5000 6.2831 13 1.000 0 0 0 0 0 F
```

278   CHAPTER 14. SOME MESH GENERATION PACKAGES

```
'MASQUE'  0 -0.4632312 1.465731 -5.0000010E-02 1.0525 0.0 0.0 0. 0 0 0  0 0 0     F
'RAYON'  0 0.2500 0. 0. 0. 0. 0. 0 0. 0 0 0     0 0 0     F
'ANGLE'  0 0.0000 0. 0. 0. 0. 0. 0 0. 0 0 0     0 0 0     F
'DISTANCE'  0 1.000 0. 0. 0. 0. 0. 0 0. 0 0 0     0 0 0     F
'RAPPORT'  0 1.000 0. 0. 0. 0. 0. 0 0. 0 0 0     0 0 0     F
'NOMBRE'  0 0.0000 0. 0. 0. 0. 0. 0 0. 0 0 0     0 0 0     F
'ECHELLE'  0 14.32432 0. 0. 0. 0. 0 0. 0 0 0     0 0 0     F
'NUREF'  0 0.0000 0. 0. 0. 0. 0. 0 0. 0 0 0     0 0 0     F
'NB_INTERVALS'  0 12.00 0. 0. 0. 0. 0. 0 0. 0 0 0     0 0 0     F
'RAISON'  0 1.000 0. 0. 0. 0. 0. 0 0. 0 0 0     0 0 0     F
'NUDSD'  0 1.000 0. 0. 0. 0. 0. 0 0. 0 0 0     0 0 0     F
```

The mesh preparation step (figure 14.2) is now completed.

Generation and storage of the mesh

1. press *EDIT_MESH* of the general menu to activate the EDIT_MESH application then type 4 <return> to select the default mesh generator options. The mesh is displayed with its meshed hole. To remove this undesirable mesh:
2. press *S_DOM* of the designation menu;
3. press *SUPPRIMER* of the EDIT_MESH menu (top);
4. press close to the meshed circle, the mesh of the hole disappears;
5. press **SAUVER** then type the requested saving type **am_fmt<return>** for example, then type the file name "carre_troue" and return. A file carre_troue.am_fmt is created;
6. press *FINIR* to quit EMC².

the file `carre_troue.am_fmt` has the following form:

```
44 60   -- nbs,nbt
    1    2   29      25   24   29
   24   31   29       3   43   30
   15   16   31      22   34   23
   26   25   30      23   34   31
   32   27   43      28   27   32
   22   33   34      21   40   38
    4   32   43      34   15   31
   33   14   34      21   33   22
    5    6   39      20   42   40
   28   32   39      41   42   19
    7    8   36      18   36   37
   28   39   35      34   14   15
   12   13   38       6   35   39
   28   35   17      43   27   26
   38   13   14      23   31   24
    7   36   17      31   16   29
   11   12   40       3   30    2
   37    8   44      30   25   29
   41   10   42      30   29    2
   44   10   41      29   16    1
   35    6    7      17   35    7
   36    8   37      17   36   18
```

14.3. EMC^2, A TWO-DIMENSIONAL MESH EDITOR

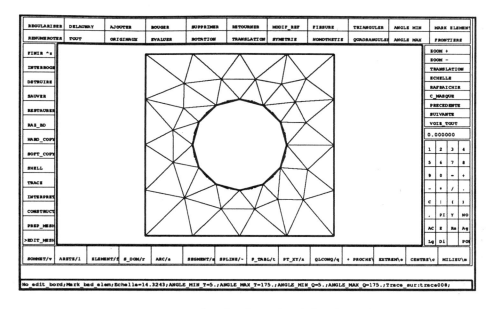

Figure 14.3: *The final mesh of the domain.*

41	37	44	18	37	19				
38	14	33	21	38	33				
39	32	4	5	39	4				
40	12	38	20	40	21				
42	10	11	37	41	19				
42	11	40	19	42	20				
43	26	30	4	43	3				
44	8	9	10	44	9				
0.000000E+00	0.000000E+00	2.500000E-01	0.000000E+00						
5.000000E-01	0.000000E+00	7.500000E-01	0.000000E+00						
1.000000E+00	0.000000E+00	1.000000E+00	2.500000E-01						
1.000000E+00	5.000000E-01	1.000000E+00	7.500000E-01						
1.000000E+00	1.000000E+00	7.500000E-01	1.000000E+00						
5.000000E-01	1.000000E+00	2.500000E-01	1.000000E+00						
0.000000E+00	1.000000E+00	0.000000E+00	7.500000E-01						
0.000000E+00	5.000000E-01	0.000000E+00	2.500000E-01						
7.500000E-01	5.000000E-01	7.165064E-01	6.250000E-01						
6.250000E-01	7.165064E-01	5.000000E-01	7.500000E-01						
3.750000E-01	7.165064E-01	2.834937E-01	6.250000E-01						
2.500000E-01	5.000001E-01	2.834936E-01	3.750001E-01						
3.749999E-01	2.834937E-01	4.999999E-01	2.500000E-01						
6.249999E-01	2.834936E-01	7.165062E-01	3.749999E-01						
2.077567E-01	2.075703E-01	4.017089E-01	1.482612E-01						
1.483122E-01	4.021604E-01	7.024530E-01	2.009149E-01						
2.021900E-01	7.009841E-01	1.471843E-01	5.790281E-01						
8.609530E-01	3.646002E-01	8.519265E-01	6.222693E-01						
7.943417E-01	7.355210E-01	2.050139E-01	8.383228E-01						
8.379703E-01	1.980158E-01	4.015485E-01	8.609668E-01						
7.210978E-01	8.353896E-01	5.831956E-01	8.603261E-01						
5.796704E-01	1.470356E-01	8.526899E-01	8.640727E-01						
0	0	0	0	0	0	0	0	0	0

o o o o o o o o o o
o o o o o o o o o o
o o o o o o o o o o
o o o o o o o o o o
o o o o o o o o o o
o o o o o o o o o o
o o o o o o o o o o
o o o o o o o o o o
o o o o o o o o o o
o o o o

Other examples of more complex geometries can be found in [Hecht,Saltel-1989].

14.4 Various other mesh generation packages

Most of the finite element packages contain a series of mesh generation processors; if not, either a mesh data structure is pre-defined and the user must create such a structure according to the required specifications, or they are coupled to software packages or C.A.D. systems which possess the required facilities.

To set up a list of the software packages with mesh generation capabilities is not an easy task. The reasons for this are that there exist a large number of software packages for mesh generation whose distribution is either local, restrained, or even confidential. On the other hand, numerous mesh generators which are more or less dedicated to particular applications exist. For these reasons, the author would welcome the contribution of any reader in order to update the following tentative list.

—— **ADINA-IN** ——

- ADINA R&D, Inc., 71 Elton Avenue, Watertown, MA 02172, USA -

—— **Araignée-Esteref**[6] ——

- LMT. ENS Cachan, 61 avenue du Président Wilson, 94230 Cachan Cedex (France) -

contact: J.P. Pelle or P. Ladevèze

—— **E3D** ——

- Cisi Ingenierie, Agence de Rungis, 3 rue Le Corbusier, SILIC 232 94528 Rungis Cedex (France) -

---

[6]2D only

14.4. VARIOUS OTHER MESH GENERATION PACKAGES

—— **Catia-FEM** ——

- Dassault Systèmes, 24-28 avenue du Général de Gaulle , B.P. 310
92156 Suresnes Cedex (France) -

contact: P. Girard

—— **Castor** ——

- CETIM, 52 avenue Félix Louat, B.P. 64 - 60304 Senlis Cedex (France) -

contact: J.M. Boissenot

—— **EMC^2** —— INRIA, BP 105 - 78153 Le Chesnay Cedex (Fr) -

contact: F. Hecht or E. Saltel[7].

—— **Euclid-IS** ——

- Matra Datavision, 31 avenue de la Baltique, B.P. 716,
Z.A. de Courtabœuf - 91961 Les Ulis Cedex (France) -

—— **Flux2d - Flux3d** ——

- INPG, rue de la Houille Blanche B.P. 46
38402 Saint Martin d'Hères (France) -

contact: J.C. Sabonnadière or J.L. Coulomb

—— **FRGEN2D-FRGEN3D** ——

- CMEE, SEAS, The George Washington University,
Washington D.C. 20052, USA -

contact: R. Löhner

—— **MacFEM**[8] ——

- Numerica, 23 Bd. de Brandenbourg, B.P. 215 - 94203 Ivry (France) -

contact: O. Pironneau

—— **MEFISTO-MAILLAGES** ——

- Laboratoire d'Analyse Numérique, Tour 55-65, Université P. et M.
Curie, 4 place Jussieu - 75252 Paris Cedex 05 (France) -

contact: A. Perronnet

—— **Mef/Mosaic** ——

---

[7] See section 14.3
[8] 2D only, suitable for Macintosh

- CSI , Technopolis , rue du Fonds Pernant - 60200 Compiègne (France) -

—— **Modulef** —— INRIA, BP 105 - 78153 Le Chesnay Cedex (Fr) -

contact: P.L. George[9]

—— **MSHPTS - MSHPTG - LGCAPA** ——

- INRIA, BP 105 - 78153 Le Chesnay Cedex (France) -

contact: A. Marrocco or F. Hecht

—— **Bacon** ——

- Samtech, 25 Bd Frère Orban, 4000 Liège (Belgium) -

contact: P. Reginster

—— **SIMAIL** ——

- Simulog, Les Quadrants, 3 avenue du centre
78182 Saint Quentin en Yvelines Cedex (France) -

contact: D. Begis

—— **I-DEAS-Supertab I-DEAS-Geomod** ——

- SDRC, 31 boulevard des Bouvets - 92000 Nanterre (France) -

—— **STRESSLAB** ——

- Prime France, Velizy Plus, 1bis rue du Petit Clamart, B.P. 25
78141 Velizy Villacoublay Cedex (France) -

contact: A. Ksiazek

—— **SWAN2D-SWAN3D-START** ——

- Computational Dynamics Research, University College Swansea,
Swansea, SA2 8PP, (U.K.) -

contact: K. Morgan or J. Peraire

—— **Systus** ——

- Framasoft, Tour Fiat, 92 Paris la Défense Cedex 16 (France) -

As a conclusion we note that the majority of commercial mesh generation software packages are structured as a series of modules which can be obtained separately. Generally, they only form a part of much larger packages containing numerous facilities, in particular at the pre and post-processing level, and furthermore have the ability to communicate with the main calculus packages (Ansys, Nastran, ...).

---

[9]See section 14.2

Chapter 15

Extensions

15.1 Introduction

A large range of topics will be discussed in this chapter. Firstly, technical details regarding the notion of a *data structure* (this concept includes two distinct meanings, as will be shown) are described, after which a practical way of entering analytical equations will be shown.

Two examples of *personal constructions* generated by the assembly of programs existing in the Modulef library are then given to emphasize the interest of the proposed mesh methodology. The first example corresponds to the programming of a preprocessor for the capturing and analysis of input data in order to construct a mesh of a composite material cell (problem of homogenization); the second example corresponds to the construction of an iterative algorithm for mesh adaptation.

The last section introduces some ideas concerning possible future developments with regard to the creation, modification and visualization of meshes.

A short conclusion ends the book.

15.2 Data structures

The storage of meshes has led to the definition of *data structures* which are composed of sets of tables. These tables contain the characteristic values of the mesh under consideration which are either directly accessible or known via some *pointers*; on the other hand, the use of a *chain* for certain values may be a very useful way of accessing them. As a general rule, to be able to perform all the manipulations necessary for a numerical or graphic process, we define an organization based on a set of *elementary data structures* amongst which the principal ones are:

- *tables*;
- *pointers*;
- *lists*;
- *stacks*;
- *queues*.

In addition, to facilitate certain computations, the techniques of:

- *chaining*;
- *coding*;
- *haching*.

can be used.

These different notions are introduced briefly in this section, after which the $NOPO$ and $GEOM$ structures, designed for mesh storage in the Modulef code, will be described in detail.

15.2.1 Basic data structures

With regard to elementary data structures, we have:

the **table**

A table TAB is a set of values which are known by a set of indices (integers). This natural definition implies that a value V is formally known by $V = TAB(i, j, k, ...)$. The indices may themselves be contained in one or several other tables.

the **pointer**

Let n be the number of sub-sets, E_i, of a given set E, then table $POINT$ defined as:

- $POINT(1) = 0$,

- $POINT(i) = \sum_{j=1}^{i-1}(cardinal(E_j))$ for $i = 2, 3, .., n+1$

15.2. DATA STRUCTURES

is a pointer for set E so that the first member of E can be accessed and then, step by step, all its elements, as $POINT(i) + 1$ is the address of the first member of sub-set i, $POINT(i+1)$ that of its last member and $POINT(i+1) - POINT(i)$ the number of members of E_i.

the list

A list $LIST$ is a table associated with a table TAB of length n which contains:

- $LIST(1)$ = address in TAB of member 1,
- $LIST(n)$ = address in TAB of the first free position (denoted "free" below)

By exploring a list, information can be *chained*, which can either be direct, inverse or multiple. This type of storage facilitates the insertion or removal operation of members in a table. As an example, the insertion of a value V as member $p+1$ corresponds to executing the following operations:

- q = free
- TAB(LIST(q)) = V
- free = LIST(free)
- LIST(q) = LIST(p)
- LIST(p) = q

We note that $LIST$ and TAB can be organized as: "LIST = tab(1,*) and TAB = tab(2,*) ".

the stack

A stack $PILE$ is a table of length n where $S = PILE(n)$ is the top of the stack and where $n = 0$ indicates that the stack is empty. A stack, associated with a process T_R, is used as follows:

- While the stack is not empty:
- Extract the top S,
- Unstack S,
- Process S and if the result of T_R induces the processing of members $e_1, e_2, ...$, stack these members.

A stack is useful in recursive processes: the solution of a problem creates several problems of the same type which are treated in sequence.

the queue

The structure of queue looks like that of the stack, but in this case the required operations are made starting at the beginning of the queue (and not at its top).

There are several ways of encoding information, among which are, in particular:

the simple encoding

Let us consider, as an illustration of this notion, a set of edges described by a table TAB with two indices which provide the number of the two endpoints of the edge. Let us assume the problem corresponds to finding (as fast as possible) an edge for which the endpoints A and B are given. A table COD is created parallel to table TAB such that: $COD(i) = function(TAB(1,i), TAB(2,i))$; this function is chosen such that the search for edge AB is accelerated. Several choices exist:

- COD(i) = first vertex of edge i,

- COD(i) = MIN(endpoint1, endpoint2),

- COD(i) = endpoint1 + endpoint2,

- ...

In order to solve the posed problem, table COD must be analyzed and, depending on its contents, index i of edge AB is found directly, or after a reduced number of computations. If the encoding is a one-to-one function, i is found immediately, otherwise, an h-coding technique can be used (see below).

the h-coding or haching

Let N be the number of members to be considered and C the number of classes. An integer value $h(x)$, called the haching value, is associated with any member x of the given set via a function h. The classes C form the main table of the haching. If box $i = h(x)$ is empty, the addresses of the C chained lists are encoded in i. As a consequence, the members of the ith list are the x's such that $h(x) = i$. To find an arbitrary member, it suffices therefore to explore only these lists. This type of haching is called

15.2. DATA STRUCTURES

open haching. A box can contain several members (in this case, there is a *collapse*).

To avoid the management of these *collapses*, a *closed haching* can be created. In the case of a box already being occupied, we look for the first free box and put the member under consideration in it. As a consequence, to find a given member x, it suffices to check if it is at position $h(x)$, if not we search through the table until it is found; if the end of the table is reached, it means that x does not exist in the present set.

It is clear that the size available influences the haching technique.

the sort

Sorting a table with respect to a given criterium is a process used frequently to facilitate the access of a value contained in a set. There are numerous sorting algorithms, the choice of which is governed by the nature of the result desired. The example below corresponds to a sort based on a *heap*:

```
      SUBROUTINE TRIEN1(CRITER,N)
C     ++++++++++++++++++++++++++++++++++++++++++++++++++
C     BUT : SORTING THE TABLE CRITER(N)
C     ---
C     ++++++++++++++++++++++++++++++++++++++++++++++++++
      INTEGER CRITER(N),L,R,J,CRIT
      IF ( N .EQ. 1 ) RETURN
      L = N / 2 + 1
      R = N
    2 IF ( L .LE. 1 ) GO TO 20
      L = L - 1
      CRIT = CRITER(L)
      GO TO 3
   20 CRIT = CRITER(R)
      CRITER(R) = CRITER(1)
      R = R - 1
      IF ( R .EQ. 1 ) GO TO 10
    3 J = L
    4 I = J
      J = 2 * J
      IF ( J - R ) 5,6,8
    5 IF ( CRITER(J) .LT. CRITER(J+1) ) J = J + 1
    6 IF ( CRIT .GE. CRITER(J) ) GO TO 8
      CRITER(I) = CRITER(J)
      GO TO 4
    8 CRITER(I) = CRIT
      GO TO 2
   10 CRITER(1) = CRIT
      END
```

Sorting a table TAB according to a criterium contained in a second table $CRITER$ is also a very useful operation for the implementation of mesh creation or modification algorithms.

The use of the structures mentioned above is constant when implementing any mesh creation algorithm, it facilitates the processes which are required: storage of information, fast search of values, updating of tables (insertion, removal, ...), and so on. In addition, the set of notions introduced will be used, on one hand, for the definition of the inner organization of the mesh storage (the example of the two Modulef mesh structures is used) and, on the other hand, for the easy manipulation of these values or derived values.

15.2.2 Data structures

The NOPO data structure

The $NOPO$ data structure is the organization chosen in the Modulef package for the mesh storage. It corresponds to an element by element description of the meshes and then provides a sequential access to the associated information. This organization is convenient with regard to the computations required during the different steps of the finite element method and, in addition, the amount of memory occupied is relatively little.

A complete description of the $NOPO$ data structure of the Modulef code is given below.

Based on the same principles as the other data structures of this code [Modulef et al. 1989], it is composed of six tables of pre-defined organization.

15.2. DATA STRUCTURES

- Table *NOP0* : General information.

    Of type integer, this table possesses 32 variables. It contains a general description of the job (title, date, name), the D.S. *NOPO* (type, level, etc.) and indicates if table *NOP1* exists.

    - 1..20 TITLE, the job title in 20 words of 4 characters each,
    - 21,22 DATE, the date in 2 words of 4 characters,
    - 23.28 NOMCRE, the name of the creator in 6 words of 4 characters,
    - 29 'NOPO', e.g. the type of structure,
    - 30 NIVEAU, the level of the D.S.,
    - 31 ETAT, one reserved parameter,
    - 32 NTACM, the number of optional supplementary tables associated with the structure (if any; they are described in table NOP1).

- Table *NOP1* : Supplementary table descriptor, if any.

    This table, of type integer, exists if NTACM (cf. *NOP0*) is non-zero. It contains, for each extra table, 22 variables. Generally, there are no supplementary tables associated with the D.S. except for certain applications when it is necessary to store values which do not enter into the classic definition, in the data structure.

    If table *NOP1* exists, it contains sequentially:

    Loop I from 1 to NTACM

    - name of table I in 4 characters,
    - address of table I, in the super table M,
    - number of words of table I,
    - type of this table (1 = integer, 2 = real*4, 4 = character, 5 = real*8, ...)
    - comments about the contents of table I in 18 words of 4 characters.

    End of loop.

- Table *NOP2* : General mesh description.

  This table, of type integer, contains 27 values.

    - 1 NDIM, the dimension of the space (2 or 3),
    - 2 NDSR, the maximum number of reference numbers,
    - 3 NDSD, the maximum number of sub-domain numbers,
    - 4 NCOPNP, the node and vertex coincidence code: 1 if they coincide, 0 otherwise,
    - 5 NE, the number of elements in the mesh,
    - 6 NEPO, the number of elements reduced to a point,
    - 7 NSEG, the number of segments,
    - 8 NTRI, the number of triangles,
    - 9 NQUA, the number of quadrilaterals,
    - 10 NTET, the number of tetrahedra,
    - 11 NPEN, the number of pentahedra,
    - 12 NHEX, the number of hexahedra,
    - 13 NSUP, the number of super-elements,
    - 14 NEF, the number of boundary elements,
    - 15 NOE, the number of nodes,
    - 16 N1, the number of nodes in a segment or on an edge (excluding endpoints),
    - 17 ISET, the number of nodes inside a triangle or a triangular face,
    - 18 ISEQ, the number of nodes inside a quadrilateral or a quadrilateral face,
    - 19 ISETE, the number of nodes inside a tetrahedron,
    - 20 ISEPE, the number of nodes inside a pentahedron,
    - 21 ISEHE, the number of nodes inside an hexahedron,
    - 22 NP, the number of points,
    - 23 NTYCOO, the type of coordinates (2 in this case, cf. *NOP1*),
    - 24 LPGDN, the maximum difference between the node numbers of an element + 1,
    - 25 NBEGM, the number of super-elements or of description in table *NOP3*,
    - 26 LNOP5, the number of words of table *NOP5*,
    - 27 NTACOO, the type of coordinate axes: 1 x, y, z; 2 r, θ, z; 3 r, θ, ϕ.

15.2. DATA STRUCTURES

- Table *NOP3* : Optional pointer.

  If NBEGM (table *NOP2*) is non-zero, this integer table describes in one variable one item of information relative to each super-element or description (this table is not currently used).

- Table *NOP4* : Vertex coordinates.

  Of type NTYCOO, this table contains the NDIM.NP vertex coordinates:

  X1, Y1, X2, Y2, ... or X1, Y1, Z1, X2, ...

- Table *NOP5* : Sequential element description.

  This table, of type integer, describes each element of the mesh, sequentially as follows:

  Loop I from 1 to NE

  - NCGE, the geometrical code of the element (1 = node, 2 = segment, 3 = triangle, 4 = quadrilateral, 5 = tetrahedron, 6 = pentahedron, 7 = hexahedron, 8 = super-element, ...)
  - NMAE, the number of words needed to store the face, edge and vertex reference numbers,
  - NDSDE, the element sub-domain number,
  - NNO, the number of nodes of the element,
    Loop J from 1 to NNO
    * NONO(J), the number of node J of the element,
    End of loop J.
  - if NCOPNP (cf. table *NOP2*) is zero:
    * NPO, the number of points of the element,
      Loop J from 1 to NPO
      · NOPO(J), the number of point J of the element,
      End of loop J.
  - if NMAE is non-zero:
    * INING, this value specifies the nature of the reference numbers stored. If INING = 3 the only numbers stored are those of the vertices, for INING = 2 the references of the edges and those of the vertices are stored and if INING = 1 the numbers stored are those of the faces, edges and vertices.

      Loop from 2 to NMAE
      · if INING = 3, the vertex references,

292   CHAPTER 15. EXTENSIONS

   · if INING = 2, the edge references and then the vertex references,
   · if INING = 1, the face, the edge and the vertex references,
   End of loop.

End of loop I.

Remark 15.1 : It is important to note that the topology of the elements is not present explicitly in table *NOP5*; see section 2.3. which describes the conventions assumed for this topology. □

Remark 15.2 : During a standard use of a D.S. *NOPO*, the user calls the modules, preprocessors, etc. of the Modulef library which automatically manage the contents of the D.S. concerned. □

Remark 15.3 : For a more refined use or when writing user programs, the reader should consult, for instance, the Modulef reports
[George et al. 1989d], [Modulef et al. 1989] in which the Modulef programming norms are detailed. □

The GEOM data structure

The *GEOM* data structure is an organization to store the information relative to a mesh in a form which is better adapted to the graphic requirements. It corresponds to a description of the elements in terms of volumes, volumes in terms of faces, faces in terms of edges and edges in terms of points. It corresponds therefore to an access of information by pointers and chains [Aho et al. 1983]. The size required for this type of structure is larger than that needed for the *NOPO* structure described above.

A complete description of the *GEOM* data structure of the Modulef code is given below.

Based on the same principles as the other structures of this code, it is composed of eight tables of pre-defined organization [Modulef et al. 1989].

- Table *GEO0* : General information.

  This table, of type integer, has 32 variables. It contains a general description of the job (title, date, name), the D.S. *GEOM* (type, level, etc.) and indicates if table *GEO1* exists.

    – 1..20 TITLE, the job title in 20 words of 4 characters each,
    – 21,22 DATE, the date in 2 words of 4 characters,

15.2. DATA STRUCTURES

- 23.28 NOMCRE, the name of the creator in 6 words of 4 characters,
- 29 'GEOM', i.e. the type of structure,
- 30 NIVEAU, the level of the D.S.,
- 31 ETAT, one reserved parameter,
- 32 NTACM, the number of optional supplementary tables associated with the structure (if any, they are described in table *GEO*1).

- Table *GEO1* : Supplementary table descriptor, if any.
  This table is analogous to table *NOP1* of the D.S. *NOPO*.

- Table *GEO2* : General mesh description.
  This table, of type integer, contains 15 values.
    - 1 NDIM, the dimension of the space (2 or 3),
    - 2 NBSOM, the number of vertices in the mesh,
    - 3 NBARET, the number of edges,
    - 4 NBFAC, the number of faces,
    - 5 NBVOL, the number of volumes,
    - 6 NTRI, the number of triangles or triangular faces,
    - 7 NQUA, the number of quadrilaterals or quadrilateral faces,
    - 8 NTET, the number of tetrahedra,
    - 9 NPEN, the number of pentahedra,
    - 10 NHEX, the number of hexahedra,
    - 11 NBSF, the number of boundary vertices,
    - 12 NBAF, the number of boundary edges,
    - 13 NBFF, the number of boundary faces,
    - 14 NCCF, the number of connected components in the domain,
    - 15 NCFF, the number of connected components on the boundary.

- Table *GEO3* : Description of the elements in terms of faces.
  This table, of type integer, contains for each element (considered as a volume):
  Loop I from 1 to NBVOL
    - NBFE, the number of faces of the element
    - Loop J from 1 to NBFE

* IPF, the address in table $GEO4$ of face J including the encoding of the orientation of its normal ($IPF > 0$ if it is an outward normal)

End of loop J.

– NDSDE, the element sub-domain number.

End of loop I.

- Table $GEO4$: Description of the faces in terms of edges.

  This integer table contains, for each face:

  Loop I from 1 to NBFAC

  – NUMGLO, the global face number,
  – NBRAR, the number of edges of face I including the encoding of its position ($NBRAR > 0$ if it is a boundary edge),
  – Loop J from 1 to NBRAR
    * IPF, the address in table $GEO5$ of edge J including the encoding of its orientation ($IPF > 0$ if anticlockwise)
    End of loop J.
  – NREFA, the face reference number.

  End of loop I.

- Table $GEO5$: Description of the edges in terms of vertices.

  This integer table contains, for each edge:

  Loop I from 1 to NBARET

  – NUM1, the number of vertex 1 of the edge,
  – NUM2, the number of vertex 2,
  – NREFR, the edge reference number.

  End of loop I.

- Table $GEO6$: Vertex coordinates.

  This real*4 table contains the vertex coordinates:

  X1, Y1, X2, Y2, ... or X1, Y1, Z1, X2, ...

- Table $GEO7$: Vertex reference numbers.

  This integer table stores the reference numbers of the NBSOM vertices.

15.2. DATA STRUCTURES

Remark 15.4 : In this structure, the definition of the points, edges and faces is slightly different to that used in the $NOPO$ data structure in the case of meshes with non-vertex nodes. For example, a "curved" edge (2 points, 3 nodes) will, in this case, be defined by a set of 2 geometrical edges (3 points, 2 segments). □

Manipulation of the mesh D.S.

Three utilization levels of the Modulef programs can be distinguished:

1. Use of preprocessors:

   In this case, the meshes created or manipulated are identified via the name of the file which contains them, however, during their manipulation in core, they are identified via an integer parameter, called a *level*, which allows the distinction between one mesh and another.

   Some of the preprocessors available are listed below:

   - APNOXX Creation and manipulation of two-dimensional meshes;
   - APN3XX Creation (for simple geometries) and manipulation of three-dimensional meshes;
   - MA23XX Mesh creation for a "cylindrical" topology;
   - RECOXX Pasting together of two meshes;
   - ......

2. Use of modules:

   In this case, the meshes under consideration are identified via the unit number associated with the file which contains them and via their *level*.

   - IMGEOM Print the contents of a structure of type $GEOM$;
   - IMNOPO Print the contents of a structure of type $NOPO$;
   - IMNOPS Selective print of values included in a D.S. $NOPO$;
   - INFOGE Information about some values contained in a D.S. $GEOM$;
   - INFONO The same in the case of a D.S. $NOPO$;
   - TRGEOM Visualization of a mesh contained in a D.S. $GEOM$;
   - TRNOPO Visualization of a mesh contained in a D.S. $NOPO$;
   - ......

3. Use of lower level subroutines:

   For this type of utilization, the reader should consult [George et al. 1989d] where the philosophy and the programming norms of the Modulef code are described and, in particular, the use of a super-table and the management of the tables associated with the D.S. are detailed.

   - SDECRI Writing the values contained in a D.S. lying in core on file;
   - SDLECT Reading a D.S. from file and storing its contents in core;
   - SDRECH Searching the tables of a D.S. in core;
   - TUERSD Removing the tables of a D.S. from core;
   - ......

15.3 Interpreted functions

15.3.1 Introduction

When creating or modifying meshes, one often has to define some functions of one or several variables. In FORTRAN, a "FUNCTION" or "SUBROUTINE" allows this kind of definition. If a module of the package requires such a function, it is necessary to:

- write the "FUNCTION" or "SUBROUTINE" associated with the function to be input;
- compile this program;
- perform the corresponding link.

Therefore, if a program calls function UTILIS, i.e., appearing as an instruction of type:

$$R = UTILIS(X,Y,Z)$$

the user must write it in the form:

FUNCTION UTILIS(X,Y,Z)
UTILIS = ...
END

Similarly, if a program calls subroutine UTILIS, i.e., if it appears as a program line of the form:

15.3. INTERPRETED FUNCTIONS

CALL UTILIS(X,Y,Z,R)

the user must write this as follows:

SUBROUTINE UTILIS(X,Y,Z,R)
R = ...
END

This solution, which seems suitable when using FORTRAN, induces several disadvantages:

- each time the user wishes to run the program with a different function, he (or she) must compile the new function and (on most of the systems) link the complete program. These operations are clearly time consuming in terms of "human" and "computer" time;

- it is not possible to change the function at the time of the execution.

To avoid these disadvantages, we have introduced the notion of interpreted functions.

15.3.2 Description of interpreted functions

See [Laug-1984] for an extensive description of these functions, of which only a brief summary follows.

The required functions are provided inside the program data. Their definition is analyzed and stored in core in an encoded form. Then, each time it is needed, each function is executed by interpreting the memorized instructions.

The lexicographical analysis is performed by a *scanner* capable of isolating and analyzing each member of the expression, i.e. :

- a numerical constant;

- an identifier;

- a unary or binary operator;

- a mathematical function;

- an open or closed parenthesis.

The type of the numerical constants is determined by an *automat with finite states* after which its value is stacked inside an integer or a real table.

Finally, all the expressions are coded by memorizing the address and type of each operand and a conventional number associated with each operator. In practice, this encoding does not respect the initial order of the elements of the expression but memorizes it in the *inverse polish* form. This technique facilitates the interpretation phase as each operand is stacked and each operator induces an operation at the top of the stack.

For example, an expression of the form:

$$A * (B + C)$$

is written as:

$$A\ B\ C\ +\ *$$

so that, at execution time, we have the following sequence:

- A, B, C are stacked,
- C and B are unstacked and the result $R1 = B + C$ is stacked,
- R1 and A are unstacked and the result $R2 = A * R1$ is stacked,

Finally, the stack contains only one member which is the final result, $R2$.

To conclude, the execution time is of course slightly higher due to the interpretation of the instructions. However, the negative consequences described above disappear: the management of the FORTRAN functions files (source and object) is suppressed, the link is done only once, and the flexibility of use is increased.

15.4 Two examples of mesh construction

15.4.1 Introduction

Two examples of programs based on the specific use of the Modulef library capabilities are given below with special attention paid to the mesh creation step. The first example concerns the design of a preprocessor facilitating data capturing corresponding to the mesh construction of cells involved when modelling a composite material. The second example deals with the design of a super-module for creating a mesh which is adapted to a given criterium for a problem in structural analysis.

15.4.2 Mesh creation for a homogenization problem

It corresponds to computing the characteristics of a homogeneous material equivalent (regarding its response) to a woven fabric; this material is composed of interlacing fibres submerged in resin. The result depends on different parameters which we adjust to obtain suitable mechanical properties.

Due to the composite nature of the fabric, the homogenization method is used. It consists of considering a cell (or period) as computational domain, which is representative of the constitution of the fabric. Only the mesh generation aspect of this elementary cell will be discussed in this section.

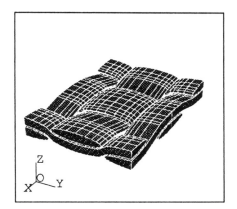
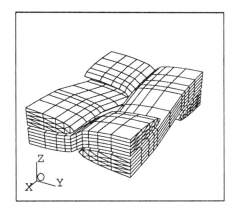

Figure 15.1: *The relative disposition of the fibres and the extracted period (by Hassim).*

The material is composed of an interlaced woven fabric submerged in a synthetic resin. At construction, this disposition is repeated: the fibres are arranged in two perpendicular directions, one direction is called the *warp* and the other the *woof*. The yarns of the woof cover one warp yarn and then go under the second woof yarn forming an interlaced pattern. This pattern is then repeated. By taking the symmetries into account, we show that only a fourth of the cell (figures 15.1 and 15.2, right-hand side) needs to be analyzed.

The parameters for defining the geometry of this cell are as follows (figure 15.3):

- the width of the yarns (warp and woof): l_x, l_y,

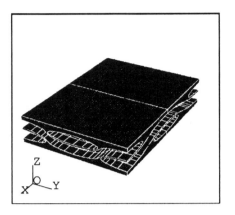 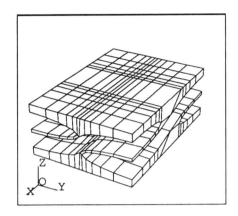

Figure 15.2: *The relative disposition of the resin and the extracted period (ibidem).*

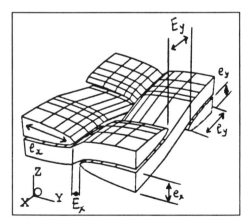 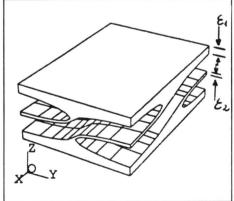

Figure 15.3: *Definition parameters.*

15.4. TWO EXAMPLES OF MESH CONSTRUCTION

- the thickness of the yarns (idem): e_x, e_y,

- the gap between two yarns (idem): E_x, E_y,

- the thickness of the resin at the crossing of yarns: ε_1,

- the thickness of the resin between two folds: ε_2.

The study consists of analyzing the effect of these values on the behaviour of the corresponding material; we will therefore vary these parameters and create, each time, the associated mesh used for the computation.

To create the mesh of the fibre and the resin, we select module COLIB2 which is a multiblock type mesh generator (cf. chapter 9). The arguments of this module are listed below (the first lines of the program are listed):

```
      SUBROUTINE COLIB2(M,NFNOPO,NINOPO,NDIMES,NBS1,NA1,NBGRO,NBFR,
     +                  XYZ,NTNRS,NARET,XYZINT,NTFR,NSUP)
C +++++++++++++++++++++++++++++++++++++++++++++++++++++++++++++++++
C BUT : MAILLAGE 2-D ET 3-D A PARTIR D'ELEMENTS GROSSIERS
C ---    ( VERSION 1989 )
C +++++++++++++++++++++++++++++++++++++++++++++++++++++++++++++++++
C IN :
C --
C M             : LE SUPER TABLEAU
C NF(NI)NOPO    : FICHIER ET NIVEAU DE LA S.D.S. NOPO
C NDIMES        : DIMENSION DE L'ESPACE ( 2 OU 3 )
C NBS1          : NOMBRE DE SOMMETS DES ELEMENTS GROSSIERS
C NA1           : NOMBRE D'ARETES DES ELEMENTS GROSSIERS
C NBGRO         : NOMBRE D'ELEMENTS GROSSIERS
C NBFR          : NOMBRE DE FACES AYANT UNE REFERENCE
C XYZ(3,NBS1)   : TABLEAU DES COORDONNEES DU MAILLAGE GROSSIER
C NTNRS(NBS1)   : TABLEAU DES NUMEROS DE REFERENCE DES SOMMETS
C NARET(5,NA1)  : TABLEAU DES ARETES DU MAILLAGE GROSSIER :
C         NARET(1,I) : SOMMET ORIGINE DE LA IEME ARETE
C         NARET(2,I) : SOMMET EXTREMITE
C         NARET(3,I) : CODE DE DECOUPAGE (0 OU 1)
C         NARET(4,I) : NOMBRE DE POINTS DE L'ARETE ( HORS EXTREMITES )
C         NARET(5,I) : NUMERO DE REFERENCE DE L'ARETE
C XYZINT(3,?)   : TABLEAU DES COORDONNEES INTERMEDIAIRES DES ARETES
C NTFR(6,NBFR)  : TABLEAU DES FACES REFERENCEES DU MAILLAGE GROSSIER :
C         NTFR(1,I)   : REFERENCE DE LA IEME FACE REFERENCEE
C         NTFR(2,I)   : TYPE DE LA FACE (3 OU 4)
C         NTFR(3-6,I) : LISTE DES SOMMETS DE LA FACE
C NSUP(11,NBGRO) : TABLEAU DES ELEMENTS GROSSIERS
C         NSUP(1,I)   : TYPE DU IEME ELEMENT GROSSIER (2 A 7)
C         NSUP(2-9,I) : LISTE DES SOMMETS DE L'ELEMENT GROSSIER
C         NSUP(10,I)  : TYPE DES ELEMENTS FINIS DE DECOUPAGE (2 A 7)
```

```
C            NSUP(11,I)  : SOUS-DOMAINE DE L'ELEMENT
C     ++++++++++++++++++++++++++++++++++++++++++++++++++++++++++++
```

The steps necessary to define the mesh data are:

- the partitioning of the domain into coarse elements (or blocks) of tetrahedral, pentahedral or hexahedral nature;

- the definition of each block by giving its vertices and a description of its edges;

- the definition of the physical attributes to take the boundary conditions, the surface loads and the two materials into account.

The set of the input values will be computed as a function of the parameters which define the geometry. A preprocessor has been designed to generate a data file for the mesh generation module from a small number of values automatically. This task represents the mesh creation step required to solve the problem under consideration [Chauchot et al. 1989]. Thus, the user only has to provide this reduced set of parameters after which the preprocessor deduces the input data required by the module from it. Figures 15.4 and 15.5 show two examples corresponding to two sets of parameters.

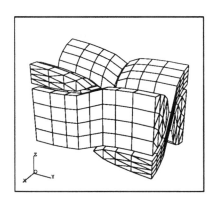 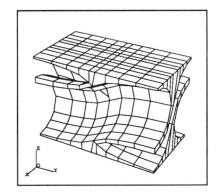

Figure 15.4: *A first geometry.*

15.4. TWO EXAMPLES OF MESH CONSTRUCTION

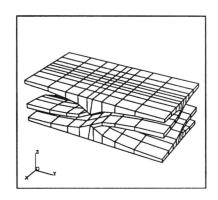

Figure 15.5: *A second geometry.*

15.4.3 An iterative algorithm for adapted mesh creation

This example concerns the integration of a mesh adaptation algorithm in the Modulef context; it was treated in a practical session in a Modulef course ([Hassim,Vidrascu-1987]).

In order to be comprehensive, this example was designed to initiate the participants of this course to the programs of the code. It is therefore not only the subject by itself which is important but also the proposed methodology which illustrates the *open feature* of the Modulef code and the consequent possibilities.

The proposed problem is the following:

- Optimize the mesh of a domain Ω by local refinement in order to solve the two-dimensional elasticity problem P_0 (described below) with plane deformations. This optimization consists of splitting the elements in such a way that a criterium, C_r, related to the constraints σ, holds.

Let us first recall the formulation of this problem and some notation:

- P_0 can be written as: Find $\underline{u} = (u_1(x,y), u_2(x,y))$ such that:

$$\begin{cases} -\sigma_{ij,j} = f_i^\Omega & i = 1,2 \text{ in } \Omega \subset R^2 \\ \sigma_{ij} n_j = f_i^\Gamma & i = 1,2 \text{ on } \Gamma_0 \\ u_i = \overline{u_i} & i = 1,2 \text{ on } \Gamma_1 \end{cases} \quad (15.1)$$

where

- $\epsilon_{ij} = 0.5(u_{i,j} + u_{j,i})$ $i, j = 1, 2$ is the strain tensor,

- $\sigma_{ij} = E_{ijkh}\epsilon_{kh}$ $i, j = 1, 2$ is the stress tensor which can be written in terms of λ and μ the Lame coefficients (for the isotropic case assumed here): $\sigma_{ij} = \lambda \epsilon_{kk} \delta_{ij} + 2\mu \epsilon_{ij}$

and:

- $\Gamma = \Gamma_0 \cup \Gamma_1$,

- n_j the jth component of the unit outward normal on Γ_0,

- f_i^Ω and f_i^Γ, the surface loads applied on Ω and line loads applied on Γ,

- $\overline{u_i}$ the imposed displacement,

- E_{ijkh} the elasticity tensor.

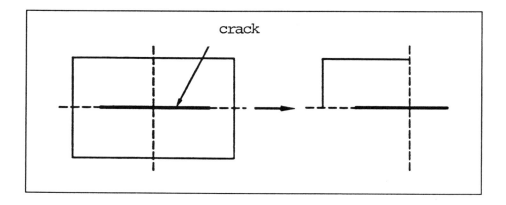

Figure 15.6: *The sample problem.*

The chosen criterion C_r, expressed in terms of the computed constraints, enables us to locate the zones where the solution varies strongly.

Therefore, let NE be the number of elements in the current mesh and NEL be the current element number, then:

$$C_r(NEL) = \frac{\sum_{i \in V(NEL)} |\sigma_i - \sigma_{NEL}|^2}{|\sigma max|^2}$$

is the criterion where:

15.4. TWO EXAMPLES OF MESH CONSTRUCTION

$V(NEL)$ is the set of the elements neighbouring element NEL,
σmax is the maximum of the constraints in domain Ω,
and $|\ |$ is a suitable norm.

An element must be refined if $C_r(NEL) \geq \alpha$ where α is given. This implies that there is a large variation in the constraints in this zone. It remains, now, to define the algorithm to go from the current state of the mesh of Ω to the new state governed by this criterium: the scheme in figure 15.7 shows an example of this process.

We will now examine the nature of the work to be done, as a function of the programs existing in the Modulef code Modulef, for each phase of this algorithm.

Let T_0 be an initial mesh and T_i the current mesh, we must:

1 compute the solution $\underline{u}^i$ and the constraints σ^i;

2 evaluate the criterion C_r;

3 modify, if necessary, mesh T_i in order to create T_{i+1}.

To compute the solutions, the following operations are required:

1.1 the definition of the interpolation in mesh T_i;

1.2 the description of the boundary conditions on Γ_1;

1.3 the computation of the element arrays (matrices and right-hand sides);

1.4 the assembly and solution of the system;

1.5 the computation of the constraints in each element using the solution now known.

To evaluate the criterion, we:

2.1 look for the elements neighbouring a given element;

2.2 compute the difference at the constraint level in order to find the list of the elements to treat;

2.3 deduce the list of nodes around which the refinement is required (cf. REGMA2, chapter 13).

Finally, the current mesh is modified by:

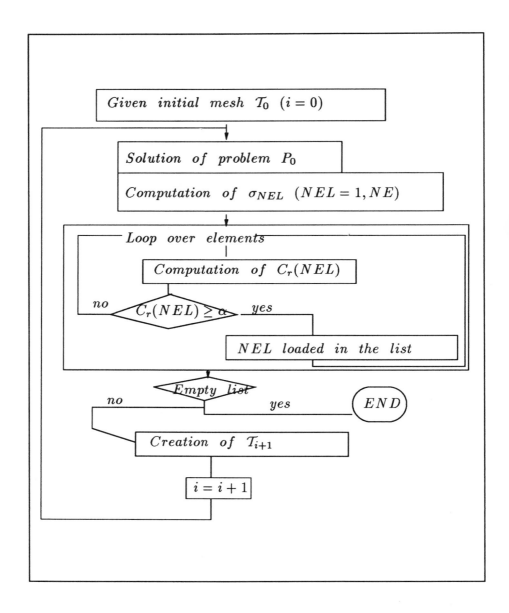

Figure 15.7: *General scheme for control and adaptation.*

15.4. TWO EXAMPLES OF MESH CONSTRUCTION

3.1 local refinement around the nodes listed in the previous step.

A program corresponds to each of these steps. The following two situations appear:

- the program exists in the Modulef code: we only need to insert it in a "super module";
- the desired program does not exist: in this case, it has to be written using the programming norms of the code in such a way that it can be inserted in the process automatically.

By consulting [Bernadou et al. 1988], the reader will see that, among the facilities already present in the code, there are programs to solve steps 1.1, 1.2, 1.3, 1.4 and 3.1, however to deal with steps 2.1, 2.2 and 2.3, the user needs to write the corresponding programs himself. Assuming this, he must concatenate the programs of the code with those of his own in such a way that the algorithm in figure 15.7 holds. Let us now discuss these two steps:

A **Writing of new programs**: to implement steps 2.1 and 2.3, we must develop the following two programs:

- "ELVOIS" to find the list of the neighbours of each element (i.e. elements sharing a vertex).
- "ELNOEU" to find the list of nodes belonging to a list of elements.

B **Writing of the general concatenation**: this task consists of two phases:

- The writing of the general module, i.e. :
    * "ADAPT2" for the computation of the solution to the elasticity problem at each iteration. It suffices to call the modules in the Modulef code which perform these types of operation.
    * "CRITER" for the analysis of the constraint values and to deduce the list of nodes around which the refinement is required. This program uses therefore the two modules ELVOIS and ELNOEU (described above).
    * "ADAPT1" the general program linking all the modules in the concatenation. This module is written using all the preceding tools: it computes the solution (ADAPT2), measures the criterium (CRITER) and then orders the refinement (AFFNOP) and, finally, initializes the next iteration.

- The writing of the "calling program" of module ADAPT1. To make it easy-to-use, this program should be conversational and as user-friendly as possible.

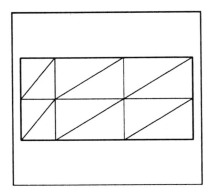

Figure 15.8: *The initial guess.*

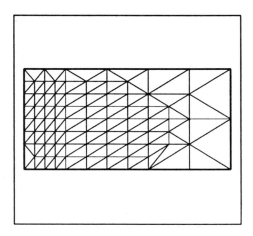
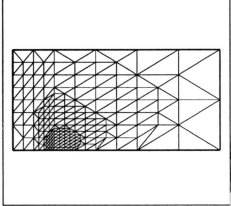

Figure 15.9: *Evolution of the mesh (2 states).*

Figure 15.8 shows the initial mesh, containing 12 elements and 12 nodes. The following plots (figure 15.9) show the meshes obtained at steps 3 and 5 of the process, containing respectively 147 and 339 elements and 87 and

15.4. TWO EXAMPLES OF MESH CONSTRUCTION

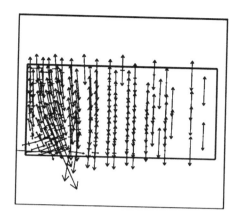
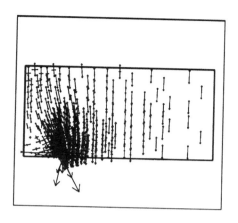

Figure 15.10: *The principal constraints.*

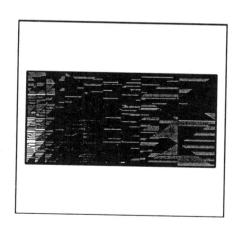
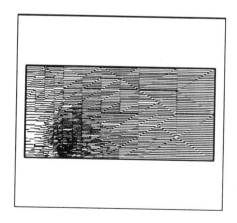

Figure 15.11: *The constraints (Tresca's criterium).*

187 nodes. The constraints corresponding to these two meshes are shown in figures 15.10 (principal constraints) and 15.11 (Tresca's criterium).

To conclude, we note that the programs to be written have a diverse nature: some are standard programs using classical **FORTRAN** tables, others are chains of programs existing in the code and programs created to manipulate the data structures of the code. The important point to notice is that, thanks to the open feature of the code, the user has access to several levels of programming and, as a consequence, is capable of implementing his own particular problems.

15.5 Future trends and conclusion

The programming norms proposed for the construction of a set of modules dedicated to the creation and the manipulation of meshes, allows for the easy and efficient evolution of the corresponding package. In particular, it is possible to improve it by modifying certain constitutive subroutines or by adding new subroutines. Considering, once more, the mesh generation part in the Modulef library, it seems reasonable (in terms of time required) to:

- modify existing modules in order to improve them;
- create new modules associated with new algorithms;
- develop user-friendly interfaces to activate only certain programs in such a way as to construct "black boxes" dedicated to this or that specific application.

Thus, this type of package can be enlarged every year with new capabilities to keep track of the progress in numerical methods (both from the methodological and algorithmic aspect) in the domain considered. In addition, the fast evolution in the performance and capabilities of computers can be taken into account when designing or programming the new modules which can be integrated *naturally* into the package, as long as its norms are respected.

In this respect, we illustrate the design of a prototype (figure 15.13) of a more attractive mesh visualization software package than that which is currently available (figure 15.12). Those who have known Modulef for several years can give account of the evolution of this code through the years (a new release being available every year).

15.5. FUTURE TRENDS AND CONCLUSION 311

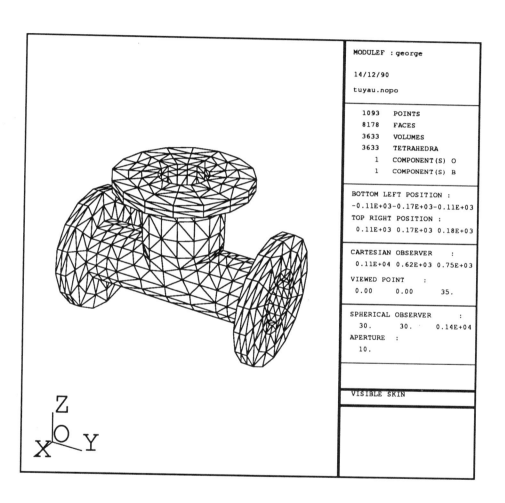

Figure 15.12: *Current module.*

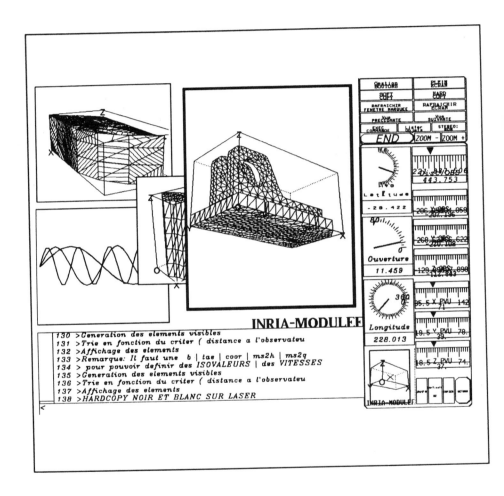

Figure 15.13: *Prototype in development.*

15.5. FUTURE TRENDS AND CONCLUSION

To conclude, the author hopes that this book has given the reader sufficient information regarding the mesh generation algorithms and the underlying methodology. He further hopes the different points which were investigated have thrown some light on the numerous difficulties connected with the mesh generation problem and will promote new approaches or lead to improvements in this field. In particular, it seems evident that studies, improvements and developments related to topics such as:

- the adaptation of meshes with respect to criteria connected to the physical properties of the problem being modelled. In this regard, the following two ideas are encountered: the creation of a mesh already enjoying all or a part of the properties required; in this case, it is the mesh generator that is designed to generate elements satisfying these properties - the modification of an existing mesh to create a new mesh possessing the desired properties.

- the development of more general algorithms to deal with the problem of the creation of the surface mesh in accordance with their description.

- the improvement (in terms of time efficiency) of the mesh generation methods, particularly a crucial point for simulations in the three-dimensional case.

- the exploitation of the capabilities of vector or parallel computers by finding algorithms taking advantage of these properties (see for example [Field,Yarnall-1989] on the implementation of a Voronoï method on a Cray X-MP). One of the main difficulties to be solved lies in the fact that most of the methods are by nature sequential (except the multiblock method).

- the integration of an automatic mesh generation process in:

    - C.A.D. software packages,
    - the solving methods of the computational codes in such a way that it will be possible to act iteratively on the mesh as a function of the solutions obtained at an iteration,
    - the coupling together of these software packages (C.A.D. codes and computational codes),
    - ...

will be developed rapidly in the very near future.

Legend

Fig. A

Top-down analysis and bottom-up construction: the domain, a homocinetic junction, is composed of three sections: the upper section, the lower section and the rubber junction. The upper section, which has a cylindrical geometry, is processed using the extrusion method (chapter 5). The mesh of the lower section is processed in the same manner while modifying certain descriptive parameters. The multiblock method (chapter 9) is applied to create the mesh of the junction. The mesh of the entire domain results of the pasting together of the meshes of the three sub-sets above.

Fig. B

Mesh of a gear: the extrusion method (chapter 5) is employed with a $2D$ mesh of the basis of the domain as input. In the first plot, visible faces of the object are shown. The second plot shows the surface of the domain in dashed line.

Fig. C

Mesh of a rotor and a stator of an electrical device. Dashed line view and visible faces of the surface of the domain are shown.

Fig. D

Mesh of a period of a composite material including a fibre and the associated matrix (chapter 15).

All these colour prints were created with the financial support of INRIA.

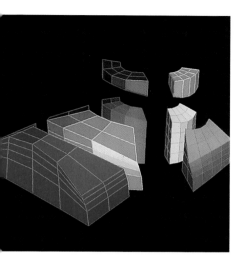 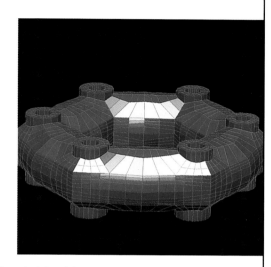

Figure 0.1: Fig. A (1) - (2)

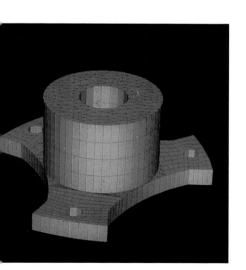 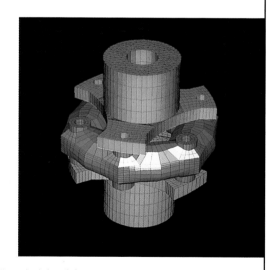

Figure 0.2: Fig. A (3) - (4)

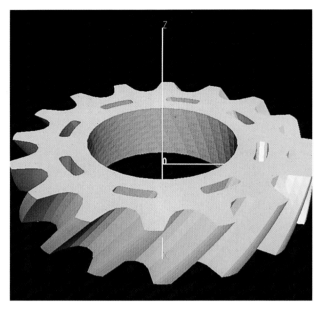

Figure 0.3: Fig. B (1)

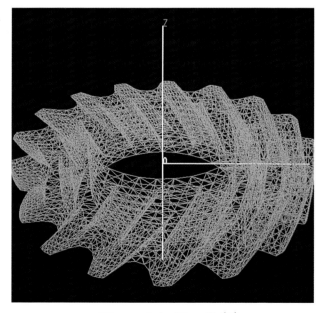

Figure 0.4: Fig. B (2)

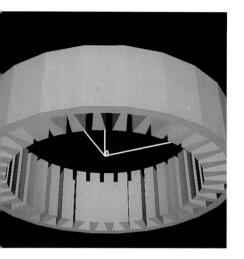 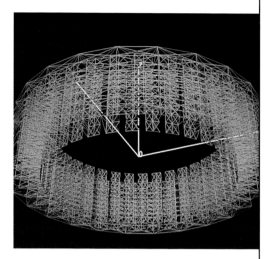

Figure 0.5: Fig. C (1) - (2)

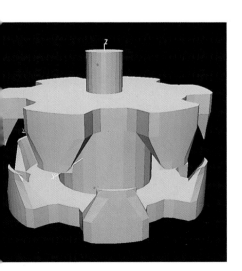 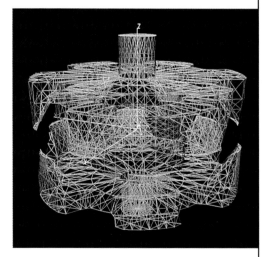

Figure 0.6: Fig. C (3) - (4)

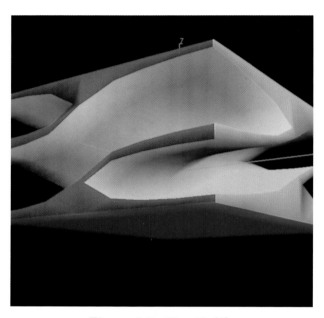

Figure 0.7: Fig. D (1)

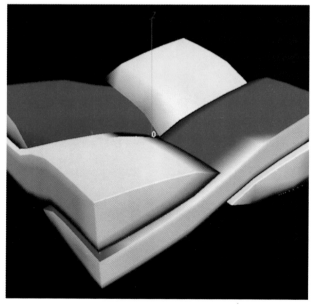

Figure 0.8: Fig. D (2)

Bibliography

[Aho et al. 1983] A. AHO, J. HOPCROFT, J. ULLMAN, Data structures and algorithms, Addison-Wesley, Reading, Mass., 1983.

[Akhras,Dhatt-1976] G. AKHRAS, G. DHATT, An automatic node relabelling scheme for minimizing a matrix or network bandwidth, Int. J. Num. Meth. Eng. 10, pp 787-797, 1976.

[Argyris-1954-1955] J.H. ARGYRIS, Energy theorems and structural analysis, part I: General theory, Aircraft Engineering 26, 1954 and 27, 1955.

[Argyris,Mlejnek-1986-1988] J.H. ARGYRIS, H.P. MLEJNEK, Die methode der finiten elemente, Band 1, 2, 3. Friedr. Vieweg and Sohn, Braunschweig/Wiesbaden, 1986-1988.

[Babuska,Aziz-1972] I. BABUSKA, A.K. AZIZ, The Mathematical Foundations of the Finite Element Method, ed. A.K. Aziz, Academic press, New York, 1972.

[Bai,Brandt-1987] D. BAI, A. BRANDT, Local mesh refinement multilevel techniques, SIAM J. of Stat. Comp., vol 8, no. 2, pp 109-134, 1987.

[Baker-1986] T.J. BAKER, Three dimensional mesh generation by triangulation of arbitrary point sets, Appl. Num. Math., vol 2, no. 1, 1986.

[Baker-1987] T.J. BAKER, Mesh generation by a sequence of transformations, Proc. AIAA 8th Comp. Fluid Dynamics Conf. Honolulu, HI, 1987.

[Baker-1988a] T.J. BAKER, Generation of tetrahedral meshes around complete aircraft, Numerical grid generation in computational fluid mechanics '88, Miami, 1988.

[Baker-1988b] T.J. BAKER, Unstructured mesh generation and its application to the calculation of flows over complete aircarft, Tennessee Univ., Space Institute Workshop, 1988.

[Baker-1989] T.J. BAKER, Developments and trends in three-dimensional mesh generation, Appl. Num. Math. 5, pp 275-309, 1989.

[Baker-1989] T.J. BAKER, Automatic Mesh Generation for Complex Three-Dimensional Regions Using a Constrained Delaunay Triangulation, Eng. Comp. vol 5, pp 161-175, 1989.

[Baker et al. 1988] B.S. BAKER, E. GROSSE, C.S. RAFFERTY, Nonobtuse triangulation of polygons, Disc. Comp. Geom., 3, pp 147-168, 1988.

[Bank et al. 1983] R.E. BANK, A.H. SHERMAN, A. WEISER, Refinement algorithms and data structure for regular local mesh refinement, Scientific Computing, R. Stepleman et al. (eds), IMACS, North Holland, 1983.

[Bartels et al. 1987] R.H. BARTELS, J.C. BEATTY, B.A. BARSKY, An Introduction to Splines for use in Computer Graphics and Geometric Modeling, Morgan Kaufmann, Los Altos, California, 1987.

[Bathe-1982] K.J. BATHE, Finite element procedures in engineering analysis, Prentice Hall, 1982.

[Bathe,Chae-1989] K.J. BATHE, S.W. CHAE, On automatic mesh construction and mesh refinement in Finite Element analysis, Computers and Structures, vol 32, no. 3/4 , pp 911-936, 1989.

[Berger-1978] M. BERGER, Géométrie tome 3 : convexes et polytopes, polyèdres réguliers, aires et volumes, Fernand Nathan, Paris, 1978.

[Berger,Oliger-1984] M.J. BERGER, J. OLIGER, Adaptive mesh refinement for hyperbolic partial differential equations, J. Comp. Phys. 53, pp 484-512, 1984.

[Bernadou et al. 1988] M. BERNADOU ET AL., Modulef: une bibliothèque modulaire d'éléments finis, INRIA, 1988.

[Bézier-1986] P. BÉZIER, Courbes et surfaces, Mathématiques et CAO, vol 4, Hermes, 1986.

[Boissonnat-1984] J.D. BOISSONNAT, Geometric structures for 3-dimensional shape representation, ACM Transactions on graphics 3, no. 4, 1984.

[Bowyer-1981] A. BOWYER, Computing Dirichlet tesselations, The Comp. J., vol 24, no. 2, pp 162-167, 1981.

[Burger-1976] M.J. BURGER, Zone a finite element mesh generator, Lawrence Livermore Lab., 1976.

BIBLIOGRAPHY

[Bykat-1976] A. BYKAT, Automatic generation of triangular grid, Int. J. Num. Meth. Eng. 10, pp 1329-1342, 1976.

[Bykat-1983] A. BYKAT, Design of a recursive, shape controlling mesh generator, Int. J. Num. Meth. Eng. 19, pp 1375-1390, 1983.

[Carnet-1978] J. CARNET, Une méthode heuristique de maillage dans le plan pour la mise en oeuvre des éléments finis, Thèse, Paris, 1978.

[Carnet,Dujardin-1987] J. CARNET, J. DUJARDIN, Maillage automatique 3D pour la mise en oeuvre de la méthode des éléments finis, le programme MATETRA, L.E.A., Le Havre, 1987.

[Caughey-1978] D.A. CAUGHEY, A systematic procedure for generating useful conformal coordinates mappings, Int. J. for Num. Meth. in Eng., vol 12, 1978.

[Cavendish-1974] J.C. CAVENDISH, Automatic Triangulation of Arbitrary Planar Domains for the Finite Element Method, Int. J. Num. Meth. Eng. 8, pp 679-696, 1974.

[Cavendish-1975] J.C. CAVENDISH, Local mesh refinement using rectangular blended finite elements, J. of Comp. Phys., vol 19, pp 211-228, 1975.

[Cavendish et al. 1985] J.C. CAVENDISH, D.A. FIELD, W.H. FREY, An approach to automatic three-dimensional finite element mesh generation, Int. J. Num. Meth. Eng. vol 21, pp 329-347, 1985.

[Cendes et al. 1985] Z.J. CENDES, D.N. SHENTON, H. SHAHNASSER, Magnetic field computations using Delaunay triangulations and complementary finite element methods, IEEE Trans. Magnetics 21, 1985.

[Cendes,Shenton-1985] Z.J. CENDES, D.N. SHENTON, Adaptive mesh refinement in the finite element computation of magnetic fields, IEEE Trans. Magnetics 21, 1985.

[Cendes,Shenton-1985a] Z.J. CENDES, D.N. SHENTON, Complementary error bounds for foolproof finite element mesh generation, North Holland, Math. and Comp. in Simulation 27, pp 295-305, 1985.

[Cendes,Shenton-1985b] Z.J. CENDES, D.N. SHENTON, Three-dimensional finite element mesh generation using Delaunay tesselation, IEEE Trans. Magnetics 21, pp 2535-2538, 1985.

[Chauchot et al. 1989] P. CHAUCHOT, O. GUILLERMIN, A. HASSIM, F. LÉNÉ, Caractérisation mécanique des structures bobinées par homogénéisation et méthodes expérimentales, in Extensiométrie appliquée aux composites, Pluralis, pp 129-141, 1989.

[Cheng et al. 1988] J.H. CHENG, P.M. FINNIGAN, A.F. HATHAWAY, A. KELA, W.J. SCHOEDER, Quadtree/octree meshing with adaptive analysis, Numerical grid generation in computational fluid mechanics'88, Miami, 1988.

[Cherfils,Hermeline-1990] C. CHERFILS, F. HERMELINE, Diagonal swap procedures and characterizations of 2D-Delaunay triangulations, Rairo, Math. Mod. and Num. Anal., vol 24, no. 5, pp 613-626, 1990.

[Chew-1989] L.P. CHEW, Constrained Delaunay Triangulations, Algorithmica, 4, 1989.

[Chew,Drysdale-1985] L.P. CHEW, R.L. DRYSDALE, Voronoi diagrams based on convex distance functions, ACM 0-89791-163-6, pp 235-244, 1985.

[Chinn et al. 1982] P.Z. CHINN, J. CHVATALOVA, A.K. DEWDNEY, N.E. GIBBS, The bandwidth problem for graphs and matrices- A survey, J. of Graph Theory, vol 6, pp 223-254, 1982.

[Ciarlet-1978] P.G. CIARLET, The Finite Element Method, North Holland, 1978.

[Ciarlet-1986] P.G. CIARLET, Elasticité Tridimensionnelle, RMA no. 1, Masson, Paris, 1986.

[Ciarlet-1988] P.G. CIARLET, Mathematical Elasticity, Vol 1: Three-Dimensional Elasticity, North-Holland, 1988.

[Ciarlet,Raviart-1972a] P.G. CIARLET, P.A. RAVIART, The combined effect of curved boundaries and numerical integration in isoparametric finite element methods in the mathematical foundations of the finite element method with applied to partial differential methods, Academic Press, pp 409-474, 1972.

[Ciarlet,Raviart-1972b] P.G. CIARLET, P.A. RAVIART, General Lagrange and Hermite interpolation in R^n with applications to finite element method, Arch. Rational Mech. Anal., vol 46, 1972.

[Clough-1960] R.W. CLOUGH, The Finite Element Method in Plane Stress Analysis, Proc. of the 2nd ASCE conf. on Electronic Computation, Pittsburg, Penn., 1960.

[Cook-1974]. W.A. COOK, Body oriented coordinates for generating 3-dimensional meshes, Int. J. Num. Meth. Eng. 8, pp 27-43, 1974.

[Coons-1967] S.A. COONS, Surfaces for computer-aided design of space forms, Technical Rep. MAC-TR 44 MIT, MA, USA, 1967.

[Coulomb-1987]   J.L. COULOMB, Maillage 2D et 3D. Experimentation de la triangulation de Delaunay, Conf. on Automated mesh generation and adaptation, Grenoble, 1987.

[Coulomb et al. 1985]   J.L. COULOMB, Y. DU TERRAIL, G. MEUNIER , Flux3d: a finite element package for magnetic computation, Proc. of Compumag'85, 1985.

[Coxeter et al. 1959]   H.S.M. COXETER, L. FEW, C.A. ROGERS, Covering space with equal spheres, Mathematika 6, pp 147-151, 1959.

[Cuthill-1972]   E. CUTHILL, Several strategies for reducing the bandwidth of sparse symmetric matrices, in Sparse matrices and their applications, Plenum Press, New York, 1972.

[Cuthill,McKee-1969]   E. CUTHILL, J. MCKEE, Reducing the bandwidth of sparse symmetric matrices, Proc. 24th Nat. Conf. Assoc. Comput. Mach., pp 157-172, 1969.

[De Boor-1972]   C. DE BOOR, On calculating with B-splines, J. Approx. theory, vol 6, pp 50-62, 1972.

[De Casteljau-1985]   P. DE CASTELJAU, Formes à pôles, Mathématiques et CAO, vol 2, Hermes, 1985.

[Delaunay-1934]   B. DELAUNAY, Sur la sphère vide, Bul. Acad. Sci. URSS, Class. Sci. Nat., pp 793-800, 1934.

[Désidéri-1990]   J.A. DÉSIDÉRI, Communication personnelle, 1990.

[Doursat,Perronnet-1989]   C. DOURSAT, A. PERRONNET, Maillage structuré. Vérification de l'injectivité et amélioration, in "Calcul des structures et intelligence artificielle", Pluralis, 1989.

[Ecer et al. 1985]   A. ECER, J.T. SPYROPOULOS, J.D. MAUL, A three-dimensional block-structured Finite Element Grid generation scheme, AIAA J., vol 23, pp 1483-1490, 1985.

[Eiseman-1979]   P.R. EISEMAN, A multi-surface method of coordinate generation, J. Comput. Phys., 33, 1979.

[Eiseman-1985]   P.R. EISEMAN, Alternating direction adaptive grid generation, AIAA J., 23, 1985.

[Eriksson-1982]   L.E. ERIKSSON, Generation of boundary-conforming grids around wing-body configurations using transfinite interpolation, AIAA J. 20, 1982.

[Eriksson-1983]   L.E. ERIKSSON, Practical three-dimensional mesh generation using transfinite interpolation, Von Karman Inst. for Fluids Dynamics, Lecture Series Notes, 1983.

[Farin-1987]   G. FARIN, Geometric Modeling: Algorithms and new trends, SIAM, 1987.

[Farin-1988]   G. FARIN, Curves and Surfaces for Computer Aided Geometric Design, Academic Press, 1988.

[Field-1987]   D.A. FIELD, Mathematical Problems in Solid Modeling: A Brief Survey, in Geometric Modeling: Algorithms and new trends, G.E. Farin, SIAM, pp 91-107, 1987.

[Field-1988]   D.A. FIELD, Laplacian smoothing and Delaunay triangulations, Comm. in Appl. Num. Meth., vol 4, pp 709-712, 1988.

[Field,Yarnall-1989]   D.A. FIELD, K. YARNALL, Three dimensional Delaunay triangulations on a Cray X-MP, in Supercomputing 88, vol 2, Science et Applications, IEEE C.S. and ACM Sigarch, 1989.

[Fiorot,Jeannin-1989]   J.C. FIOROT, P. JEANNIN, Courbes et Surfaces Rationnelles: Applications à la CAO, RMA no. 12, Masson, Paris, 1989.

[Flux2d]   FLUX2D, Cedrat-Zirst, 38240 Meylan France.

[Fritz-1987]   W. FRITZ, Two-dimensional and three-dimensional block structured grid generation techniques, in Conf. on automated mesh generation and adpation, Grenoble, France, 1987.

[Garey et al. 1978a]   M.R. GAREY, D.S. JOHNSON, F.P. PREPARATA, R.E. TARJAN, Triangulating a simple polygon, Inform. Proc. Letters 7, 1978.

[Garey et al. 1978b]   M.R. GAREY, R.I. GRAHAM, D.S. JOHNSON, D.E. KNUTH, Complexity results for bandwidth minimization, SIAM J. Appl. Math., vol 34, no. 3, pp 477-495, 1978.

[Gaudel et al. 1987]   M.C. GAUDEL, M. SORIA, C. FROIDEVAUX, Types de Données et Algorithmes : Recherche, Tri, Algorithmes sur les Graphes, collection didactique, no. 4, vol 2, INRIA, 1987.

[Geomod]   GEOMOD, User's manual, SDRC Inc., Milford, Ohio, 1986.

[A. George-1971]   J.A. GEORGE, Computer implementation of the finite element method, Stan-CS, Ph. D., 1971.

[A. George-1973]   J.A. GEORGE, Nested dissection of a regular finite element mesh, SIAM J. Num. Anal., vol 10, pp 1345-367, 1973.

[George,Liu-1978]   J.A. GEORGE, J.W. LIU, An automatic nested dissection algorithm for irregular finite element problems, SIAM J. Num. Anal., vol 15, pp 1053-1069, 1978.

[A.George,Liu-1981] J.A. GEORGE, J.W. LIU, Computer solution of large sparse positive definite systems, Prentice Hall, 1981.

[George-1988] P.L. GEORGE, Modulef : Génération automatique de maillages, collection didactique no. 5, 2nd édition, INRIA, 1988.

[George-1989a] P.L. GEORGE, Modulef : Construction et modification de maillages, Rapport Technique no. 104, INRIA, 1989.

[George-1989b] P.L. GEORGE, Modulef : Mesh generation (user's guide), Rapport Modulef no. 104, INRIA, 1989.

[George-1989c] P.L. GEORGE, La conception descendante appliquée à la réalisation concrète des maillages, Rapport Technique no. 107, INRIA, 1989.

[George-1989d] P.L. GEORGE, Mailleur 3D par découpage structuré d'éléments grossiers, Rapport de Recherche no. 990, INRIA, 1989.

[George-1990] P.L. GEORGE, Génération des maillages pour la simulation par éléments finis de problèmes physiques, in "Revue de physique appliquée", vol 25, no. 7, pp 567-581, 1990.

[George,Golgolab-1988] P.L. GEORGE, A. GOLGOLAB, Mailleur 3D en topologie "cylindrique", Rapport Technique no. 100, INRIA, 1988.

[George et al. 1987] P.L. GEORGE, A. GOLGOLAB, B. MULLER, E. SALTEL, Bibliothèque Modulef: aspects graphiques, Rapport Modulef no. 96, INRIA, 1987.

[George et al. 1988] P.L. GEORGE, F. HECHT, E. SALTEL, Tétraédrisation automatique et respect de la frontière, Rapport de Recherche no. 835, INRIA, 1988.

[George et al. 1988] P.L. GEORGE, F. HECHT, E. SALTEL, Constraint of the boundary and automatic mesh generation, in "Numerical grid generation in computational fluid mechanics", Miami, 1988.

[George et al. 1988] P.L. GEORGE, F. HECHT, E. SALTEL, Mailleur automatique en tétraèdres respectant une frontière donnée, in "Calcul des structures et intelligence artificielle", Pluralis, 1988.

[George et al. 1989a] P.L. GEORGE, F. HECHT, E. SALTEL, Maillages automatiques de domaines tridimensionnels quelconques, Rapport de Recherche no. 1021, INRIA, 1989.

[George et al. 1989] P.L. GEORGE, F. HECHT, E. SALTEL, Automatic triangulations by a pseudo-Voronoï technique in 3D, workshop on "Computational Mathematics and Applications", Pavie, Italie, 1989.

[George et al. 1989b] P.L. GEORGE, F. HECHT, E. SALTEL, Fully automatic mesh generator for 3D domains of any shape, Impact of Computing in Science and Engineering, vol 2, n° 3, pp 187-218, 1990.

[George et al. 1989d] P.L. GEORGE, P. LAUG, B. MULLER, M. VIDRASCU, Guide d'utilisation et normes de programmation, Rapport Modulef no. 1, INRIA, 1989.

[George et al. 1990a] P.L. GEORGE, F. HECHT, E. SALTEL, Automatic mesh generation with specified boundary, to appear in Comp. Meth. in Appl. Mech. and Eng., 1990.

[George et al. 1990b] P.L. GEORGE, F. HECHT, E. SALTEL, Automatic 3D mesh generation with prescribed meshed boundaries, IEEE Trans. Magnetics, vol 26, no. 2, pp 771-774, 1990.

[George,Hermeline-1989] P.L. GEORGE, F. HERMELINE, Maillage de Delaunay d'un polyèdre convexe en dimension d. Extension à un polyèdre quelconque, Rapport de Recherche no. 969, INRIA, 1989.

[George,Hermeline-1989] P.L. GEORGE, F. HERMELINE, Delaunay's mesh of a convex polyhedron in dimension d. Application for arbitrary polyhedra, submitted to Int. J. Num. Meth. Eng., 1990.

[Gibbs et al. 1976] N.E. GIBBS, W.G. POOLE, P.K. STOCKMEYER, An algorithm for reducing the bandwidth and profile of a sparse matrix, SIAM J. Num. Anal. 13, no. 2, pp 236-250, 1976.

[Gibbs et al. 1976] N.E. GIBBS, W.G. POOLE, P.K. STOCKMEYER, A comparison of several bandwidth and profile reduction algorithm, ACM Trans. on Math. Software, 2, pp 322-330, 1976.

[Glowinski-1984] R. GLOWINSKI, Numerical Method for Nonlinear Variational Problems, Springer Verlag, 1984.

[Golgolab-1989] A. GOLGOLAB, Mailleur tridimensionnel automatique pour des géométries complexes, Rapport de Recherche no. 1004, INRIA, 1989.

[Gordon,Hall-1973] W.J. GORDON, C.A. HALL, Construction of curvilinear coordinate systems and applications to mesh generation, Int. J. Num. Meth. Eng. 7, pp 461-477, 1973.

[Grice et al. 1988] K.R. GRICE, M.S. SHEPHARD, C.M. GRAICHEN, Automatic, topologically correct, three-dimensional mesh generation by the finite octree technique, RPI center for interactive Computer graphics, 1988.

[Grooms-1972] H.R. GROOMS, Algorithm for matrix bandwidth reduction, J. of Structural division, vol 98, no. ST1, pp 203-214, 1972.

[Hacon,Tomei-1989] D. HACON, C. TOMEI, Tetraedral decompositions of hexahedral meshes, Europ. J. Combinatorics, vol 10, pp 435-443, 1989.

[Hall-1976] C.A. HALL, Transfinite interpolation and applications to engineering problems, in Law and Sahney, Theory of approximation, Academic Press, pp 308-331, 1976.

[Hackbush,Trottenberg-1982] W. HACKBUSCH, U. TROTTENBERG, Multigrid methods, Lect. notes in Mathematics 960, Springer Verlag, 1982.

[Hassim,Vidrascu-1987] A. HASSIM, M. VIDRASCU, Intégration dans Modulef d'un algorithme de maillage adaptatif, Rapport Interne Modulef, INRIA, 1987.

[Hauser,Taylor-1986] J. HAUSER, C. TAYLOR, Numerical grid generation in computational fluid dynamics, Pinebridge Press, 1986.

[Hecht,Saltel-1989] F. HECHT, E. SALTEL, Emc2 : Un logiciel d'édition de maillages et de contours bidimensionnels, Rapport Technique no. 118, INRIA, 1990.

[Hermeline-1980] F. HERMELINE, Une méthode automatique de maillage en dimension n, Thèse, Université Paris 6, Paris, 1980.

[Hermeline-1982] F. HERMELINE, Triangulation automatique d'un polyèdre en dimension N, Rairo, Analyse numérique, vol 16, no. 3, pp 211-242, 1982.

[Ho Le-1988] K. HO LE, Finite element mesh generation methods: a review and classification, Comp. Aided Design, vol 20, pp 27-38, 1988.

[Holmes,Snyder-1988] D.G. HOLMES, D.D. SNYDER, The generation of unstructured triangular meshes using Delaunay triangulation, Numerical grid generation in computational fluid mechanics'88, Miami, 1988.

[Hubert] G. HUBERT, Mosaic: Stratégie et algorithmes de maillage, Compiègne Science Industrie, 60200 Compiègne.

[Hughes-1988] T.J.R. HUGHES, The Finite Element Method, Prentice-Hall Int., 1988.

[norme-a] IGES, Initial graphic exchange standard, Ansi.

[Jacquotte-1988] O.P. JACQUOTTE, A mechanical model for a new generation method in computational fluid mechanics, Comp. Meths. Appl. Mech. Engrg., vol 66, pp 323-338, 1988.

[Jameson et al. 1984] A. JAMESON, T.J. BAKER, N.P. WEATHERILL, Calculation of inviscid transonic flow over a complete aircraft, Proc. AIAA, 24th Aerospace Sciences Meeting, Reno, NV, 1984.

[Joly-1988] P. JOLY, Mise en oeuvre de la méthode des Elements Finis, Univ. P. et M. Curie, Laboratoire d'Aanlyse Numérique, no. 88002, Paris, 1988.

[Johnson-1987] C. JOHNSON, Numerical Solution of P.D.E. by the Finite Element Method, Cambridge University Press, 1987.

[Joukowsli-1910] N. JOUKOWSKI, Ueber die konturen der drachen flieger, ZFM, 1910.

[Lascaux,Theodor-1986] P. LASCAUX, R. THEODOR, Analyse Numérique Matricielle Appliquée à l'art de l'ingénieur, Masson, Paris, 1986.

[Laug-1984] P. LAUG, Les fonctions interpretées: Manuel d'utilisation, de programmation, de référence, Rapport Technique no. 38, INRIA, 1984.

[Lawson-1972] C.L. LAWSON, Generation of a triangular grid with application to contour plotting, California institute of Technology, JPL, 299, 1972.

[Lawson-1986] C.L. LAWSON, Properties of n-dimensional triangulations, Computer Aided Geometric Design, 3, pp 231-246, 1986.

[Lee-1980] D.T. LEE, Two-dimensional Voronoi diagrams in Lp-Metric, J. of the ACM, vol 27, no. 4, pp 604-618, 1980.

[Lee et al. 1980] K.D. LEE, M. HUANG, N.J. YU, P.E. RUBBERT, Grid generation for general three-dimensional configurations, Proceedings NASA Langley Workshop on Numerical Grid Generation Techniques , 1980.

[Lee,Schachter-1980] D.T. LEE, B.J. SCHACHTER, Two algorithms for constructing a Delaunay Triangulation, Int. J. Comp. Inf. Sci., Vol 9, no. 3, 1980.

[Lipton,Tarjan-1979] R.J. LIPTON, R.E. TARJAN, A separator theorem for planar graphs, SIAM J. Appl. Math., vol 36, no. 2, pp 177-189, 1979.

[Lipton et al. 1979] R.J. LIPTON, D.J. ROSE, R.E. TARJAN, Generalised nested dissection, SIAM J. Appl. Math., vol 16, no. 2, pp 346-358, 1979.

[Lo-1985] S.H. LO, A new mesh generation scheme for arbitrary planar domains, Int. J. Num. Meth. Eng. 21, pp 1403-1426, 1985.

[Lo-1989]　　　S.H. Lo, Delaunay triangulation of non-convex planar domains, Int. J. Num. Meth. Eng. 28, pp 2695-2707, 1989.

[Löhner-1988]　　R. LÖHNER, Some useful Data Structures for the Generation of unstructured grids, Comm. Appl. Num. Meth. no. 4, pp 123-135, 1988.

[Löhner,Parikh-1988a]　R. LÖHNER, P. PARIKH, Three-dimensional grid generation by the advancing-front method, Int. J. Num. Meth. Fluids, no. 8, pp 1135-1149, 1988.

[Löhner,Parikh-1988b]　R. LÖHNER, P. PARIKH, Generation of 3D unstructured grids by the advancing front method, AIAA 88 0515, 26th Aerospace Sciences meeting, Reno, NV, 1988.

[Marro-1980]　L. MARRO, Méthodes de réduction de largeur de bande et de profil efficace des matrices creuses, Thèse, Université de Nice, 1980.

[Marrocco-1984]　A. MARROCCO, Simulations numériques dans la fabrication des circuits à semiconducteurs (process modeling), Rapport de Recherche no. 305, INRIA, 1984.

[Meagher-1982]　D. MEAGHER, Geometric modelling using octree encoding, Comp. Graph. Image Proc., vol 19, pp 129-147, 1982.

[Melhem-1987]　R. MELHEM, Towards efficient implementation of preconditioned conjugate gradient methods on vector supercomputers, Int. Jour. of Supercomp. Appl., vol 1, 1987.

[Milgram-1989]　M.S. MILGRAM, Does a point lie inside a polygon? J. of Comp. Phys., vol 84, pp 134-144, 1989.

[Modulef et al. 1989]　MODULEF ET AL., Description des Structures de Données, Rapport Modulef no. 2, INRIA, 1989.

[Newman,Sproull-1979]　W.M. NEWMAN, R.F. SPROULL, Principles of interactive computer graphics, McGraw-Hill Kogakusha, 1979.

[Oden-1972]　J.T. ODEN, Finite Element of Nonlinear Continua, McGraw-Hill, 1972.

[Oden,Carey-1984]　J.T. ODEN, G.F. CAREY, The Texas Finite Element Series, Prentice-Hall, 1984.

[Palmerio-1987]　B. PALMERIO, A two-dimensional F.E.M. adaptive moving node method for steady Euler flow simulation, Publication de l'université de Nice no. 150, 1987.

[Palmerio-1988] B. PALMERIO, A consistent ale-rezoned mesh adaption algorithm for compressive flow finite-element calculations, Rapport de Recherche no. 829, INRIA, 1988.

[norme-b] PDES, Product description exchange standard, ISO.

[Peraire et al. 1987] J. PERAIRE, M. VAHDATI, K. MORGAN, O.C. ZIENKIEWICZ, Adaptive remeshing for compressible flow computations, J. of Comp. Phys., 72, no. 2, pp 449-466, 1987.

[Peraire et al. 1988] J. PERAIRE, J. PEIRO, L. FORMAGGIA, K. MORGAN, O.C. ZIENKIEWICZ, Finite element Euler computations in three dimensions, Int. J. Num. Meth. in Eng. vol 26, pp 2135-2159, 1988.

[Perronnet-1974] A. PERRONNET, Le module APMEFI, IRIA, 1974.

[Perronnet-1985] A. PERRONNET, Cours de DEA, Univ. Paris 6, 1985.

[Perronnet-1988a] A. PERRONNET, Tétraèdrisation d'un objet multimatériaux ou de l'extérieur d'un objet, Laboratoire d'analyse numérique 189, Université Paris 6, 1988.

[Perronnet-1988b] A. PERRONNET, A generator of tetrahedral finite elements for multi-material object and fluids, Numerical grid generation in computational fluid mechanics'88, Miami, 1988.

[Perronnet et al. 1977] A. PERRONNET, P. JOLY, O. KOUTCHMY, Les modules de maillage bidimensionnel du Club Modulef, IRIA, 1977.

[Peruchio et al. 1989] R. PERUCCHIO, M. SAXENA, A. KELA, Automatic mesh generation from solid models based on recursive spatial decompositions, Int. J. for Num. Meth. in Eng., vol 28, pp 2469-2501, 1989.

[Pierrot et al. 1979] R. PIERROT, J. VAZEILLES, A. PERRONNET, Une méthode de génération d'un maillage bi ou tridimensionnel à partir d'un maillage grossier, Annales de l'Institut Technique du Bâtiment et des Travaux Publics, no. 372, 1979.

[Potter,Tuttle-1973] D.E. POTTER, G.H. TUTTLE, The construction of discrete orthogonal coordinates, J. of Computational Physics, vol 13, 1973.

[Preparata,Hong-1977] F.P. PREPARATA, S.J. HONG, Convex hull of finite sets of points in two and three dimension, Com. of the ACM, vol 20, no. 2, 1977.

[Preparata,Shamos-1985] F.P. PREPARATA, M.I. SHAMOS, Computational geometry, an introduction, Springer-Verlag, 1985.

[Raiser et al. 1989] A. RAISER, G.MEUNIER, J.L. COULOMB, An approach for automatic adaptive mesh refinement in finite element computation of magnetic fields, IEEE trans. on Magnetics, vol 22, no. 4, pp 2965-2968, 1989

[Raviart,Thomas-1983] P.A. RAVIART, J.M. THOMAS, Introduction à l'Analyse Numérique des Equations aux Dérivées Partielles, Masson, Paris, 1983.

[Risler-1989] J.J. RISLER, Méthodes mathématiques pour la C.A.O., RMA n° 18, Masson, Paris, 1991.

[Rivara-1984a] M.C. RIVARA, Mesh refinement processes based on the generalised bisection of simplices, SIAM J of Num. Ana., vol 21, pp 604-613, 1984.

[Rivara-1984b] M.C. RIVARA, Algorithms for refining triangular grids suitable for adaptive and multigrid techniques, Int. J. of Num. Meth. Eng., vol 21, pp 745-756, 1984.

[Rivara-1986] M.C. RIVARA, Adaptive finite element refinement and fully irregular and conforming triangulations, in Accuracy Estimates and Adaptive Refinements in Finite Element Computations, I. Babuska et al., John Wiley and Sons, 1986.

[Rivara-1990] M.C. RIVARA, Selective refinement/derefinement algorithms for sequences of nested triangulations, Int. J. of Num. Meth. Eng., vol 28, no. 12, pp 2889-2906, 1990.

[Roman-1980] J. ROMAN, Sur la renumérotation des noeuds d'interpolation d'un maillage plan d'éléments finis à l'aide de l'algorithme de séparation de Lipton et Tarjan, Thèse, Université de Bordeaux 1, 1980.

[Sabonnadiere,Coulomb-1986] J.C. SABONNADIÈRE, J.L. COULOMB, Eléments finis et CAO, Hermes, Paris, 1986.

[Sadek-1980] E.A. SADEK, A scheme for the automatic generation of triangular finite elements, Int. J. Num. Meth. Engng., 15, pp 1813-1822, 1980.

[Saltzman-1986] J. SALTZMAN, Variational methods for generating meshes on surfaces in three dimensions, J. of Comp. Phys., 63, pp 1-19, 1986.

[Samcef] SAMCEF, Système pour l'analyse des milieux continus par la méthode des éléments finis, Manuel théorique LTAS, Liège.

[Samet-1984] H. SAMET, The quadtree and related hierarchical data structures, Computing Surveys, vol 16, no. 2, pp 187-285, 1984 [norme-c] SET, Standards d'échange et de transferts, Aerospatiale.

[Shephard et al. 1988] M.S. SHEPHARD, F. GUERINONI, J.E. FLAHERTY, R.A. LUDWIG, P.L. BAEHMANN, Finite octree mesh generation for automated adaptive 3D flow analysis, Numerical grid generation in computational fluid mechanics '88, Miami, 1988.

[Shephard et al. 1988a] M.S. SHEPHARD, F. GUERINONI, J.E. FLAHERTY, R.A. LUDWIG, P.L. BAEHMANN, Adaptive solutions of the Euler equations using finite quadtree and octree grids, Computers and Structures, 30, pp 327-336, 1988.

[Shephard et al. 1988b] M.S. SHEPHARD, K.R. GRICE, J.A. LO, W.J. SCHROEDER, Trends in automatic three-dimensional mesh generation, Comp. and Struct., vol 30, no. 1/2, pp 421-429, 1988.

[Shepard,Yerri-1982] M.S. SHEPHARD, M.A. YERRI, An approach to automatic finite element mesh generation, Comp. in Eng., vol 3, pp 21-28, 1982.

[Shpitalni et al. 1989] M. SHPITALNI, P. BAR-YOSEPH, Y. KRIMBERG, Finite element mesh generation via switching function representation, Finite element in Analysis and design, vol 5, pp 119-130, 1989.

[Sloan-1987] S.W. SLOAN, A fast algorithm for constructing Delaunay triangulations in the plane, Adv. Eng. Soft. vol 9, no. 1, pp 34-55, 1987.

[Sloan,Houlsby-1984] S.W. SLOAN, G.T. HOULSBY, An implementation of Watson's algorithm for computing 2-dimensional Delaunay triangulations, Adv. Eng. Soft. vol 6, no. 4, pp 192-197, 1984.

[Steger,Sorenson-1980] J.L. STEGER, R.L. SORENSON, Use of hyperbolic partial differential equations to generate body-fitted coordinates, Numerical Grid Generation, Nasa Conf. pub., CP-2166, 1980.

[Steinhoff-1986] J. STEINHOFF, Blending method for grid generation, J. of Comp. Physics, vol 65, 1986.

[Strang,Fix-1973] G. STRANG, G.J. FIX, Analysis of the Finite Element Methods, Prentice Hall, 1973.

[Supertab] SUPERTAB, User's manual, SDRC Inc., Milford, Ohio, 1986.

[Talon-1987a] J.Y. TALON, Algorithmes de génération et d'amélioration de maillages en 2D, Rapport Technique no. 20, Artemis-Imag, 1987.

[Talon-1987b] J.Y. TALON, Algorithmes d'amélioration de maillages tétraèdriques en 3 dimensions, Rapport Technique no. 25, Artemis-Imag, 1987.

[Tarjan,Van Wyk-1988] R.E. TARJAN, C. VAN WYK, An O(nloglog n)-time algorithm for triangulating a simple polygon, SIAM, J. of Comp., 17, pp 143-177, 1988.

[Thacker-1980] W.C. THACKER, A brief review of techniques for generating irregular computational grids, Int. J. Num. Meth. Engng., 15, pp 1335-1341, 1980.

[Thacker et al. 1980] W.C. THACKER, A. GONZALEZ, G.E. PUTLAND, A method for automating the construction of irregular computational grids for storm surge forecast models, J. of Comp. Phys., 37, pp 371-387, 1980.

[Thompson-1982a] J.F. THOMPSON, Numerical grids generation, Appl. Math. and Comp., vol 10-11, 1982.

[Thompson-1982b] J.F. THOMPSON, Elliptic grids generation, numerical grids generation, Elsevier Science, pp 79-105, 1982.

[Thompson-1985] J.F. THOMPSON, Z.U.A. WARSI, C.W. MASTIN, Numerical grids generation, foundations and applications, North Holland, 1985.

[Thompson-1987] J.F. THOMPSON, A general three dimensional elliptic grid generation system on a composite block-structure, Comp. Meth. Appl. Mech. and Eng., vol 64, 1987.

[Triboix-1985] A. TRIBOIX, Une méthode numérique inverse pour l'équation de Laplace. Applications aux écoulements et à la génération automatique de maillage, J. de Mécanique théorique, 4, 2, 1985.

[Vallet-1990] M.G. VALLET, Génération de maillages anisotropes adaptés - Application à la capture de couches limites, Rapport de Recherche no. 1360, INRIA, 1990.

[Voronoi-1908] G. VORONOÏ, Nouvelles applications des paramètres continus à la théorie des formes quadratiques. Recherches sur les parallélloedres primitifs. Journal Reine angew. Math. vol 134, 1908.

[Vouillon] X. VOUILLON, Titus, Spie-Batignolles, Paris.

[Watson-1981] D.F. WATSON, Computing the n-dimensional Delaunay Tesselation with applications to Voronoï polytopes, Computer Journal 24, no. 2, pp 167-172, 1981.

[Weatherill-1985] N.P. WEATHERILL, The generation of unstructured grids using Dirichlet tesselation, MAE report no. 1715, Princeton Univ. 1985.

[Weatherill-1988] N.P. WEATHERILL, A method for generating irregular computational grids in multiply connected planar domains, Int. J. Num. Meth. Fluids, 8, 1988.

[Weatherill-1988] N.P. WEATHERILL, A strategy for the use of hybrid structured-unstructured meshes in CFD, Num. Meth. for Fluids Dynamics, Oxford University Press, 1988.

[Weatherill-1990] N.P. WEATHERILL, The integrity of geometrical boundaries in the 2-dimensional Delaunay triangulation, Comm. in Appl. Num. Meth., vol 6, pp 101-109, 1990.

[Winslow-1964a] A.M. WINSLOW, An irregular triangle mesh generator, R. no. UCRL-7880, NTIS, Springfield, VA, 1964.

[Winslow-1964b] A.M. WINSLOW, Equi-potential zoning for two-dimensional meshes, R. no. UCRL-7312, 1964.

[Winslow-1967] A.M. WINSLOW, Numerical solution of the quasilinear Poisson equation in a non uniform triangle mesh, J. Comp. Phys., 1, 1967

[Wordenweber-1984] B. WORDENWEBER, Finite element mesh generation, Computer-aided-design, vol 16, pp 285-291, 1984.

[Yerri,Shephard-1983] M.A. YERRI, M.S. SHEPHARD, A modified quadtree approah to finite element mesh generation, IEEE CG-A, 1983.

[Yerri,Shephard-1984] M.A. YERRI, M.S. SHEPHARD, Automatic 3D mesh generation by the modified-octree technic, Int. Jour. Num. Meth. Eng., vol 20, pp 1965-1990, 1984.

[Yerri,Shephard-1985] M.A. YERRI, M.S. SHEPHARD, Automatic mesh generation for three-dimensional solids, Comp. Struct. 20, 1985

[Zienkiewicz-1971] O.C. ZIENKIEWICZ, The Finite Element Method in Engineering Science, McGraw-Hill, 1971.

[Zienkiewicz-1973] O.C. ZIENKIEWICZ, Finite Element. The background story, The Mathematics of Finite Element and Applications, Academic Press, 1973.

[Zienkiewicz-1977] O.C. ZIENKIEWICZ, The Finite Element Method, McGraw-Hill, London, 1977.

[Zienkiewicz,Phillips-1971] O.C. ZIENKIEWICZ, D.V. PHILLIPS, An automatic mesh generation scheme for plane and curved surfaces by isoparametric co-ordinates, Int. Jour. Num. Meth. Eng., vol 3, pp 519-528, 1971.

[Zlamal-1973] M. ZLAMAL, Curved elements in the finite element method, SIAM J. in Numer. Anal., vol 10, 1973.

Index

A
adaptation   215, 236, 303
advancing front method   32, 137, 138, 147, 223
Aitkens   194
algebraic method   77
anisotropic   7, 217
assembly   5
attribute   14, 39

B
B-Spline   202
barycentrage   234
beam   25
Bernstein   196, 205
Bezier   188, 197, 201
bisection   228
bottom-up construction   29, 46, 52
boundary condition   5, 14, 26, 39, 40

C
C-type grid   109
C.A.D.   52, 187, 190
Cardinal-Spline   202
Catmull-Rom   199, 201
characteristic line   53
characteristic point   53
collapse   147, 221
combination method   32
computational geometry   186
conformal mesh   5
conformity   5

connectivity   8, 14, 19, 109, 210
constraint   223
control space   155, 184, 238
criterion of the empty sphere   163
cubic geometry   70
curve   191
cylindrical geometry   72

D
Data Structure   16
De Casteljau   197
deformation method   31, 77, 129, 232
Delaunay   32
Delaunay mesh   159, 164
Delaunay-Voronoï method   32
departure zone   137
derefinement   17
dilation   217

E
element density   7, 31
elliptical method   99
encoding   286
envelope   164, 169, 171, 172
error estimation   7

F
face   57
finite element method   2
front   137, 139, 148

G
generating system   98
GEOM   16, 17, 292, 293

GKS 257
graphic 257
grid 8, 111

H
h-coding or haching 220
h-method 237
H-type grid 109
haching 286
Hermite 196, 198, 200
hexahedron 24
hinge 25
history 13, 14
hybrid multiblock method 123
hyperbolic method 106

I
IGES 16
internal constraint 222
interpolation 13
interpreted functions 297
isotropic 7, 217
item 51, 53, 54

J
junction 25

L
Lagrange 20, 193, 200
line 56
list 285
local remesh 228

M
manual method 30, 65
mesh generation method 9
membrane problem 2
mesh 4, 5
mesh transformation 215
method based on the solution of P.D.E. 97
method by solution of P.D.E. 31
method of deformation 90
methods for creation of meshes 30
modeller 9
multiblock method 32, 123
multigrid 238

N
neighbourhood space 154
node 14
non-obtuse mesh 7
NOPO 16, 17, 288
numbering 15, 20, 22, 25, 222

O
O-type grid 109
octal tree 114
octree 111, 120
overlapping method 31

P
p-method 242
P1 element 20
P2 element 20
P3 element 20
painter's algorithm 260
parabolic method 106
parallel computer 313
pasting 126, 215, 220
pasting of two meshes 37
patch 205, 209, 209
PDES 16
pentahedron 24
PHIGS 257
point 55, 191
pointer 125, 131, 285
primal hybrid triangle 20
primal sub-set 35, 51

Q
quadrilateral 23
quadtree 111, 115
quality 1, 169, 170, 180
quaternary tree 114
queue 286

R

INDEX

r-method   237
reference   15, 26, 39, 43
refinement   17, 226
regularization   119, 122, 236
renumbering   215, 245
respect of a boundary   165, 169
rotation   217

S

secondary sub-set   35
segment   22
semi-automatic method   30, 65
SET   16
simplex   16
sort   287
stack   285
star-shaped set   161
structured mesh   8
structured partitioning method   31
sub-domain   15, 25, 41
symmetry   216

T

table   284
tetrahedron   23
three-noded triangle   20
top-down analysis   29, 34
topology   14, 19, 25
transfinite interpolation   84
translation   217
transport-mapping method   31, 77
triangle   22
triangulation   5

U

unstructured mesh   8

V

variational formulation   2
vector computer   254, 313
visualization   215, 254
volume   58
Voronoï method   159, 224

MASSON Éditeur
120, bd Saint-Germain
75280 Paris Cedex 06
Dépôt légal : novembre 1991

3355 - La Bayeusaine graphique
6-12, rue royale, 14401 Bayeux
Dépôt légal : 8256
Octobre 1991